AF701816

GREEN STRUCTURES

Energy Efficiency in Buildings

GREEN STRUCTURES

Energy Efficiency in Buildings

S. Ghosh

A. Dhaka

CRC Press
Taylor & Francis Group
Boca Raton London New York

CRC Press is an imprint of the
Taylor & Francis Group, an **informa** business
A TAYLOR & FRANCIS BOOK

Green Structures: Energy Efficiency in Buildings

S Ghosh, A Dhaka

First published 2016 by Ane Books for CRC Press

Published 2024 by CRC Press
2385 NW Executive Center Drive, Suite 320, Boca Raton FL 33431

and by CRC Press
4 Park Square, Milton Park, Abingdon, Oxon, OX14 4RN

First Edition 2016

CRC Press is an imprint of Taylor & Francis Group, LLC

© 2016 Author

Reasonable efforts have been made to publish reliable data and information, but the author and publisher cannot assume responsibility for the validity of all materials or the consequences of their use. The authors and publishers have attempted to trace the copyright holders of all material reproduced in this publication and apologize to copyright holders if permission to publish in this form has not been obtained. If any copyright material has not been acknowledged please write and let us know so we may rectify in any future reprint.

Except as permitted under U.S. Copyright Law, no part of this book may be reprinted, reproduced, transmitted, or utilized in any form by any electronic, mechanical, or other means, now known or hereafter invented, including photocopying, microfilming, and recording, or in any information storage or retrieval system, without written permission from the publishers.

Trademark notice: Product or corporate names may be trademarks or registered trademarks and are used only for identification and explanation without intent to infringe.

British Library Cataloguing in Publication Data
A catalogue record for this book is available from the British Library

ISBN: 978-1-4987-4963-3 (hbk)

Visit the Taylor & Francis Web site at
http://www.taylorandfrancis.com

and the CRC Press Web site at
http://www.crcpress.com

For distribution in rest of the world other than the Indian sub-continent

Foreword

The world population is forecast to continue to grow. At the same time economic growth is expected to continue across the globe, with industrial expansion being most rapid in the emerging economies of Asia, most notably China and India. These two factors present a major challenge to conservation of finite reserves of fossil fuels. The combustion of these fossil fuels and subsequent release of carbon dioxide is known to have already resulted in significant changes to global climate. Both the future conservation of fossil fuels and the need to restrain climate change mean that the major economies of the world must be committed to alternative technologies to generate power by renewable sources. Simultaneously, techniques should be employed which use energy in the most efficient way possible.

The authors have a rich overseas experience in this area enabling them to have a broad perspective from which to view global issues of energy consumption. Their book considers the design of energy efficient buildings with a specific focus on India. The text is well supported by photographs to illustrate practical applications of design concepts. The bridge between our past and future homes and work places is effectively made by considering both traditional and futuristic designs of buildings. The book considers both fundamental scientific principles on energy conservation as well as drawing on more complex disciplines such as the flow of air around buildings.

The book is divided into six chapters. Chapter 1 on Energy Efficiency in Buildings considers the historical perspective of building design and the environmental impacts of air pollution as well as introducing the global requirement for reduced energy consumption and the principles behind low energy design. Chapter 2 is on Siting and Microclimate. This section is well supported by a number of interesting photographs and diagrams of buildings in India. It deals with the design of buildings in the context of their local climate and considers impacts of temperature, wind speed and precipitation and landscape. Chapter 3 considers Building Form and how this can be used to influence both heat loss and gain as well as lighting in historic and modern buildings. Chapter 4 is about Building Fabric and focuses on how the use of different materials can be used to control heat transfer through the basic processes of convection, conduction and radiation. The Chapter 5 deals with Renewable Energy and is split into two main sections on solar energy and wind energy. The first section considers passive solar heating, solar water heating and photovoltaic cells and the second covers atmospheric science leading to wind speed and direction profiles and the efficiency of wind turbines to generate energy. This chapter also includes harvesting biomass as fuel, rainwater collection and use of waste in landfill sites to generate methane as fuel. Chapter 6 sets the previous chapters into a global perspective by considering greenhouse gases and global warming. This chapter also covers energy efficiency ratings, auditing of energy use and building rating systems.

The book is recommended reading both for students of environmental sciences and structural design. The book presents a careful balance between scientific theory and practical application as well as a rich assortment of photographic examples and an easily readable narrative. It will therefore also have an appeal to members of the general public interested in environmental issues and energy conservation.

Dr. Tony Dore

Centre for Ecology and Hydrology, U.K.

Preface

The world is beginning to feel the crunch of depleting conventional energy sources and these desperate times call for exploring newer vistas in terms of alternative sources of energy. The current economic growth in populous developing countries, such as India and China, will exacerbate energy use in buildings. Energy Efficient Buildings are not entirely new concepts in these ancient civilizations. Our ancients used natural resources and had a deep understanding of wind flow patterns and optimal siting practices as is evident from the Sthapatya Veda, Vaastu Shaastra and Buddhist literature from India and elsewhere. This acquired wisdom has to relate to extant technology to curb the unregulated and resource intensive energy use patterns of the present day. A sporadic surge in commercialization coupled with a fierce demand for energy have led China and India to being ranked among the top 5 in the world in terms of energy consumption. If a double figure GDP has to be sustained, then the construction sector would require a major overhaul involving strategic advancements in the utilization of green energy. Given the rapid urban growth, the number of buildings as urban units is projected to grow spirally in the coming three decades. For example, the average electricity consumption for space conditioning and lighting in India is around 80kWh/m^2/ann. for residential buildings and double that for commercial buildings. But this ratio is going to change as more people move into urban conurbations. Under a business-as-usual scenario, and based on a 10% annual increase in new built-up area, the projected annual increase in electricity demand in the two sectors would be 5.4 billion kWh (TERI). The IPCC (Intergovernmental panel on climate change) Fourth assessment report reiterates the necessity of taking appropriate measures to bring down carbon emissions from the building sector.

A number of policy initiatives to mainstream energy efficiency are in place in most countries, including India. The Energy Conservation Building Code (2007) is one such example. This act led to the formation of the Bureau of Energy Efficiency (BEE).

Building rating systems such as, LEED (Leadership in Energy and Environmental Design) and GRIHA (Green Rating for Integrated Habitat Assessment) are expected to accelerate energy efficiency and sustainability in buildings. Streamlining the green building movement requires coordinated monitoring and intelligent control of a building's vital functions. This can be achieved through Building Management Systems (BMS). It involves the integration of structures, systems, services and management. Green building movements should examine the potential of power generation from urban, municipal and industrial wastes. Equitable distribution of water resources should be carefully planned–monsoon dominated countries should adopt efficient rain-water harvesting measures to provide clean water to thirsty nations.

All these overarching concepts are now well woven into themed chapters in this book. It should be easily comprehensible to undergraduates in civil/structural and environmental engineering, graduate research students, town planners and architects worldwide.

The content not only comes from taught courses on Energy Efficient Buildings, but also includes new and original research undertaken by the authors. This definitive resource book should have an enduring appeal for all concerned with green structures.

The contents of the book grew out of a taught course entitled 'Energy Efficient Buildings' which is regularly taught to B.Tech as well as M.Tech students in almost all major engineering Institutes the world over. The authors are grateful to the Chancellor, VIT University and the Dean of the School of Mechanical and Building Sciences, VIT University for their encouragement and support. Prof. Ghosh wishes to thank Prof. Lord Julian Hunt, FRS, for inspiration–the many years of active research at Cambridge University in Environmental Fluid Mechanics was possible because of Prof. Hunt's active research team. He is grateful to Prof. John Pyle, FRS, also from Cambridge University, Prof. Peter Jonas and Prof. Michael Smith from Manchester University; Dr Alan Gadian and Dr. Steve Dobbie from Leeds. Both the authors are grateful to Dr Tony Dore from the Centre for Ecology and Hydrology, Edinburgh for writing the Foreword. Prof. Ghosh wishes to thank Mr Kaushiek Pranoo, a former student who studied 'Energy in the Built Environment' for his editorial assistance for some of the chapters. In addition, discussions with Kaushiek helped to pitch the right tone for Indian B.Tech undergraduate students.

Abhinav Dhaka wishes to thank Prof. Ghosh for giving him this opportunity and being a life-infusing beacon that he is. He also wishes to express his gratitude to Dr. DVS Bhagvanulu from Gyan Ganga Institute of Technology and Sciences for unbridled support and encouragement in the years of his academic path. He is grateful to Dr. Emilia Van Egmond De Wilde De Ligny from Technical University of Eindhoven, for her guidance and input which set the precursor for this book. He also wishes to thank Dr. B R Phanikumar and Dr S Biswas for inspiring him and giving him more than an education. He wishes to express his gratitude towards his parents and his wife for it was not for their unbridled love and acceptance, the labour wouldn't have come to fruition.

The book has been brought out beautifully by ANE Publishers–the authors are most grateful to the ANE team for bringing it out so well. The way it has been brought out serves a dual purpose–it covers all the basic tenets of Green buildings whilst affording a coffee table feel to any interested reader. The book is replete with photographs of many energy efficient buildings from different parts of the world and the main inspiration for writing this textbook comes from our many visits to some exceptional buildings in the Sri Aurobindo Ashram and Auroville. These institutions cherish the ideals of energy efficiency in the built environment and were conceived by the Mother of Sri Aurobindo Ashram, Mme. Mirra Alfassa. Prof. Ghosh is especially grateful to Nirodbaran (Sri Aurobindo Ashram), Barbara (Matrimandir), Shraddhavan and Helmut (Savitri Bhavan) and Satprem (Auroville Earth Institute) for many interesting discussions and for letting me photograph these buildings.

S Ghosh, A Dhaka

Contents

1

The Essence of Energy Efficiency

Chapter Overview

1.1 Overview and Historical Perspective

Efficient use of energy has become vital since global energy resources are known to be finite. Power requirement (which mainly uses fossil fuels such as coal and oil) is radically on the rise and has adverse environmental effects. Buildings are one of the largest energy consumers, and this sector accounts for 25-40% of the net energy demand in many countries. Hence the potential for energy savings in the building sector is large. One way or another, there is a need for urgent action to be taken to reduce energy use in buildings, thus addressing climate change and energy use issues.

There are three main approaches towards this goal: reducing the energy demand for buildings (*e.g.* use of energy efficient devices), local production of energy from renewable and other wasted energy resources and buildings that can generate surplus energy and feed it into an intelligent grid infrastructure, thereby sharing energy[5].

Since the time when humans lived in caves and enjoyed the benefits of stable temperatures and natural ventilation with zero mortgage and environmental impact, we have been refining our use of resources to provide improved shelter. Fossil fuel reserves are limited and human population continues to grow, however the saving grace is that new technology and new construction materials continue to grow and help us solve building problems. However, many of these methods are not sustainable and therefore deplete our planet's resources and release by-products that are detrimental to our planet's overall well-being. Even with astonishing technical advances, our basic housing needs have generally remained unchanged, although we expect higher levels of comfort with evolving technology. In the last few thousand years, the human species has evolved physically, only marginally – however the progression of technology changed by leaps and bounds over the past three millennia. Building practices concomitantly changed, not always for the better.

The new mantra is to go green to minimize environmental impacts and develop greener technologies for building comfortable homes.

This building philosophy is not a 20th century invention – ancient India's vernacular architecture during the Upanishadic times alluded to the concepts of sustainability. The ancient Romans indigenously built green-houses and their solar heated baths and much like the provisions made by the constitution of India, the Justinian code of law during the 6th century AD promulgated an ordinance that legalized the availability of sunshine for every roman citizen. Pompeii homes had courtyards with high thermal masses, supporting a roof garden. In addition, they had adjustable sunshade devices. Today we talk of green roofs – these date back to thousands of years. The Hanging Gardens of Babylon in present day Iraq were the first examples of an intensive green roof built in deep soil. In contrast, European Vikings centuries later built green roofs that were classed as extensive and supported only a thin soil. Reports of earth sheltered buildings are now evident within mainland China and these buildings were used for centuries.[6]

In India too, the building design and technology have always been inclusive of the natural environment. The Indus Valley civilization and later Buddhist architecture show a great deal of sensitivity in resource management and energy conservation–they included efficient drainage systems, granaries, clustered dwellings and communal kitchens.

Since the primary occupation of the majority was outdoor activities like farming, *etc*, buildings served more as homes and rest shelters than workspaces. Moreover, extraordinary facilities such as natural ventilation, lighting, thermal insulation *etc*, were incorporated into the construction of these buildings right from the times of the Indus valley civilization.

1.2 Energy and Environment

Life has existed on planet Earth for about four billion years. In that time, climate has swung between ice ages and warm periods. But generally, the Earth's atmosphere has been in chemical balance, its composition changing slowly in response to changes in geology and life on the planet's surface. Now, growing population and the by-products of civilization are upsetting this balance.

Data from weather station records that Earth's surface temperature has increased on average by 0.6°C in the last 100 years. The 1990s have been the warmest decade on record. Over the same period, global sea level has increased by 10-20 cm. We know also that planet Earth has an atmosphere that creates a natural greenhouse effect, keeping our surface warmer than it would otherwise be. Human activities are substantially increasing the atmospheric concentrations of greenhouse gases, principally carbon dioxide and methane, to levels far above those that have existed for the past 200,000 years.

One of the primary reasons for this is our rapid demand for power. For the first time in history, human activity is contributing heavily to global climate change and has taken it to levels as never before with respect to global temperature rise and its consequences. With the tremendous growth of infrastructure that the country is witnessing, energy efficiency in buildings assumes paramount importance. The energy saving potential can be as high as 40-50%, if addressed right at the design stage[1].

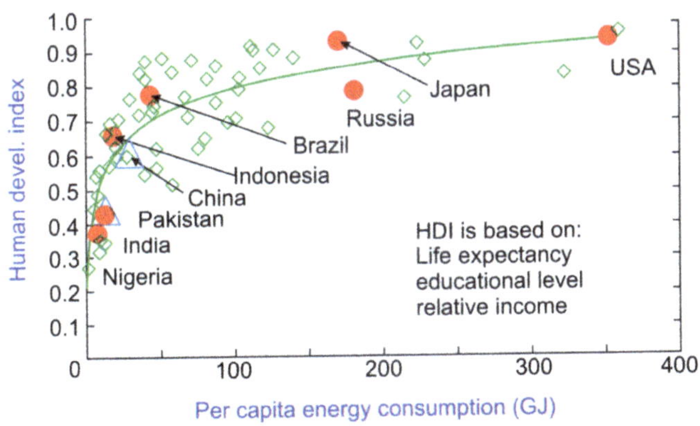

Fig. 1.1: HDI vs Per capita consumption
(Courtesy: UNDP Human Development Index Document)

Figure 1 shows that the developing nations appear on the steeply rising portion of the curve–this implies that even small adjustments to per capita energy consumption patterns can result in large changes in the Human Development Index (HDI).

1.2.1 Types of Energy

Energy is primarily the ability to do work. It manifests itself in various forms and causes a change in the properties of whatever it interacts with. Mechanical energy is the energy that causes change in the position of an object whereas radiant energy, also known as electromagnetic energy includes energy from gamma rays, x-rays, ultraviolet, visible and infrared rays. The energy released from chemical bonds during chemical reactions is called chemical energy while the heat energy of an object determines how active its atoms are. Other notable forms of energy are electrical energy caused by the drift of electrons and nuclear energy (energy stored in the nucleus of an atom), etc.[1].

1.2.2 Environmental Pollution

The use of natural resources at a rate higher than nature's capacity to replenish itself results in pollution of air, water, and land. The inability of the environment to process and neutralize harmful by-products of human activities (for example, gaseous emissions of pollutants) in due course without any structural or functional damage to its system causes environmental pollution. This matters a lot because it has negative impacts on crucial environmental services such as provision of clean air and clean water without which life on Earth as we know it would not exist.

The major types of pollution are air, water and soil pollution. Some of the most important air pollutants are sulfur dioxide, nitrogen dioxide, carbon monoxide, ozone, volatile organic compounds (VOCs) and airborne particles. Water pollutants include, food processing waste, pollutants from livestock operations, VOCs, heavy metals, insecticides and herbicides, chemical waste and others. Some soil pollutants are hydrocarbons, solvents and heavy metals.

Pollution has serious detrimental effects on the environment and living beings. In human beings it not only causes physiological disabilities but can also lead to severe psychological and neurological disorders. Acid rain (which occurs when pollutant gases dissolve in rainwater turning into acids) apart from destroying the aquatic life in lakes and streams, can also corrode metals, damage surfaces of buildings and monuments, and cause soil acidification. Water Pollution may cause oxygen depletion in marine environments and severely affect the health of whole ecosystems. The disposal of non-biodegradable wastes, including containers, bottles and cans made of plastic, used cars and electronic goods, leads to the pollution of land.

1.2.3 Reducing Energy Consumption

It is imperative that we unleash our engineering, economic, and political enterprise to improve energy conservation and efficiency and move towards greater use of renewable energy sources. Technology and training in energy efficiency and use of non-polluting fuels can allow developing nations to skip the carbon intensive, energy-production stage of industrialization.

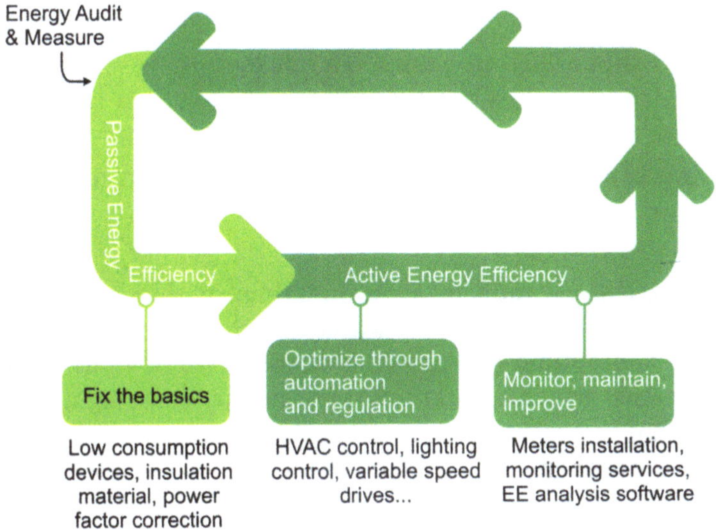

Fig. 1.2: Implementing energy efficiency to reduce energy consumption (Courtesy: www.maintenanceworld.com)

Industrial facilities are increasingly going "green"–this is good news for the environment. With the increase in energy prices, it is more likely that many sustainability initiatives in the industrial market are being driven by companies' bottom-lines. In many cases, internal sustainability projects won't get senior management approval based solely on a positive environmental impact. Going green very often must mean saving green, and this is possible when you successfully implement energy efficiency initiatives that reduce energy consumption.

A good first step in any energy optimization project is to perform a comprehensive energy analysis by examining the energy usage patterns and demands of a facility and identifying opportunities to improve energy efficiency. Often requiring little or no investment, a focus should be placed on possible behavioral changes such as shutting down production equipment and lighting when not in use to reduce energy [2]. An outline of this scheme is shown in Figure 1.2.

One of the key technologies, whose focus is on ensuring the maximum usage of the secured resources like water, air and solar energy, is green building technology. The basic principle of this technology is to ensure the maximum use of the natural resources and create less pressure on the resources that create energy for us. That is why the focus of energy experts has been on the use of renewable resources like water from the rain and the solar energy. This technology aims to eliminate the artificial pressure that is being created on the nature and to cause least amount of side effect on nature. The design of these green houses is such that it harmonizes the ecological and the visual relation between the nearest natural resources and the structure of the building. Every part of a green building must use natural resources as much as possible so that the energy segment of the environment isn't stretched to the limit.[3].

Energy conservation must be ingrained within a nation's psyche – it is an individual as well as a collective effort. On an individual level, the appropriate conservation methods are obvious – unplugging and switching off equipment when not in use, ensuring an appropriate utilization of natural resources and using only energy efficient equipment although they are marginally more expensive. Home appliances such as washing machines and air-conditioners should be used with discretion – excessive water amounts and undesirable cooling loads can be easily regulated.

This brings us to the overarching theme of energy management, individually as well as collectively. Individual attempts must compliment a global need to save energy. In the final analysis, this global need regulates not only energy prices but also affects emission targets. Appropriate legislation must be in place whenever there is a compelling reason to save energy. There are a plethora of options for saving energy – the first requirement, as alluded earlier, is assuming a sense of responsibility within oneself followed by a collectively planned effort on energy conservation.

1.2.4 Basic Principles of Low Energy Design

The principles of low-energy design are simple and essentially involve the following: the reduction in energy demand and meeting the remaining energy requirement as efficiently and cleanly as possible.

By following a conscious design process, it is possible to construct buildings that use substantially less energy without compromising occupant comfort or the building's functionality. Whole-building design considers the energy-related impacts and interactions of all building components, including the building site; its envelope (walls, windows, doors, and roof); its heating, ventilation, and air-conditioning (HVAC) system; and its lighting, controls, and equipment. Figure 1.2 outlines a streamlining of these processes.

Design mechanisms are in place that ensure low energy consumption, modern software are increasingly being used as enhanced architectural features that make intelligent use of day-lighting. Many of these design strategies involve features that act as superior climate moderators – a conventional flat façade or a flat roof have been replaced by contoured surfaces to modulate incoming solar radiation. Other examples include microclimatic adaptations around built areas with the use of specially selected trees and shrubs. A forested canopy adjacent to low-lying buildings can modulate the ingress of incoming winds through pressure differences. Water features absorb large amounts of radiation and also facilitate evaporative cooling. The building form and orientation vis-à-vis the adjacent open space can allow for better ventilation rates. In addition, building enhancements can both increase as well as decrease heat gains and losses according to the occupant's needs.

The focus should be on user comfort and their activities in the design of the thermal indoor climate such as zoning in areas with the same thermal comfort requirements and zoning in areas with the same time of occupancy and user pattern. Focus is also required on the contribution that can be gained from passive solar heating – however, it is best avoided in areas with risk of over-heated temperatures and to use it only for areas with a heating demand. Another way is to minimize the internal heat load (persons, devices and lighting) in areas with risk of over-heated temperatures.

References

1. http://www.tropical-rainforest-animals.com/Types-of-Pollution.html
2. http://www.maintenanceworld.com/Articles/impo/Energy-Optimization.html - How to turn your facility green By Cassie Quaintance, Energy Market Segment Manager, Schneider Electric posted 1-26-09
3. http://ezinearticles.com/?Reducing-Energy-Consumption&id=3175983 -http://EzineArticles.com/?expert=Md._Mahmudul_Hasan
4. Low-Energy Building Design Guidelines Energy-efficient design for new Federal facilities. DOE/EE-0249, Federal Energy Management Program, http://www.eren.doe.gov/femp/
5. http://www.pewclimate.org/docUploads/EEB Summary Report FINAL.pdf
6. http://www.yourhome.gov.au/technical/fs12.html.
7. United Nations Development Program HDI document.

2

Siting and Microclimate

Chapter Overview

Buildings set in conducive environments are not only aesthetically appealing–they are also more comfortable to live in. The building fabric (or material) can moderate the inside temperature by making it cooler (or adding to the coolth) or warmer as compared to the ambient exterior temperature. The fabric however must be carefully chosen keeping in mind the subtleties of the exterior microclimate.

2.1 Climate and Energy within the Indian Context

Climatological processes arising from variations in energy deposition on the earth by the sun dictate the patterns, of weather that we encounter. India hosts six major climatic subtypes, ranging from desert in the west, to alpine, tundra and glaciers in the north, to humid tropical regions supporting rain forests in the southwest and the island territories. Many regions have starkly different microclimates. Although the Tropic of Cancer, the boundary between the tropics and subtropics, passes through the middle of India, the whole country can be considered to be tropical. The variation in solar energy deposition repeats annually. As a consequence the climatic variables, including temperature and wind speed and direction, solar intensity and humidity can be anticipated each year, on a national scale within a reasonable degree of accuracy.

Achieving thermal comfort within a residential complex comprising of multifarious activities can be a daunting task. Nevertheless, 18th century houses in the four southern states of peninsular India were at least partially successful in achieving a good degree of thermal comfort simply by optimizing on the available daylight, the ingress of winds and a modulated microclimate. The architects of Dakshinachitra, essentially replicated a custom built housing pavilion showcasing the best forms of architecture from Tamil Nadu, Andhra Pradesh, Kerala and Karnataka.

The energy used to heat or cool a building does not remain in the building but is lost via two routes–Fabric heat loss arises when heat is transferred from the warm exterior to the cold interior through the external surfaces of the building. The reverse can also happen in tropical and warm climes. This occurs through a combination of heat transfer mechanisms: conduction through masonry, convection across air spaces such as cavities, and radiation across external masonry surface. Ventilation heat losses occur when the warm/cold air inside leaves and is replaced by cool/warm air from outside.

2.2 Macro and Microclimate

We have seen that large-scale climatic variations are modulated by the sun's energy. In addition, local effects, including topographical variations also affect wind and temperature distribution patterns. High rise buildings and other constructions can modify the local climate by drainage flows, wind channeling and overshadowing effects. Adapting buildings to the microclimate is challenging but achievable. This latter aspect has been achieved remarkably successfully in the international city of Auroville, located in the southern state of Tamil Nadu in India. Since it is located within the tropical belt, the overarching emphasis is on maintaining the building coolth. This is achieved by adding water features, trees and shrubs around the building.

2.3 Nuances of Siting

It is possible to minimize the effect of climate on buildings with today's technology and resources. The variation of climate with location, *i.e.*, site knowledge through local monitoring or meteorological office data enables one to appreciate changes in temperature and humidity– the two most important parameters required to adjust comfort levels inside a building. Secondly, it is important to understand the effect of climate on heat loss and gain. Wind, rain, solar intensity and external temperature all have a direct effect on a building's heat loss or gain rate. An understanding of climate related heat loss/gain processes is necessary for suggesting methods for preventing such losses and gains. Other factors include design of landscape to minimize the effects of climate and the reinforcement of the building fabric and form so that it acts as a climate moderator. Indeed, landscape architecture has come of age in most countries, including India.

2.3.1 Orientation

2.3.1.1 Daylighting

Natural daylight ought to be the most preferred form of illumination in buildings. Weather variations mainly modulated by the cloud cover and the passage of the sun overhead cause changes in the received patterns of daylight. Unlike electric lighting, daylight, the most magnificent of all nature's bounty is available in abundant measure in sunny countries; including India–it does not require external forms of energy to create it. However, daylight is not entirely free. It is provided through glazing systems which are thermally poor and so there is a cost in terms of additional space heating or cooling energy consumption. This energy consumption is not solely due to provision of daylighting, but is also shared by the other functions–windows also provide views and ventilate. Daylight can substantially reduce a building's energy consumption if electric lights are switched off for much of daytime work. We show in Figure 2.1a an example of a naturally lit corridor in Dakshinachitra's main entrance which has used a moderately dark paved material for its floors for aesthetic elegance. The use of dark paved material enhances the absorption of heat in the general vicinity of a building. Conduction of this heat to the surrounding air increases air temperatures. It must be remembered, however, that dark paving materials, due to their reduced daylight reflectiveness, may result in diminished daylight entry into the building–this was the desired effect required in Dakshinachitra–the diffused undertones of the tropical light lent in an aura of calm and coolth. This is in contrast to the white flooring that is used in an art gallery in Auroville's Savitri Bhavan (see Figure 2.1d) which requires reflected light from the floor for enhanced visual appeal of the wall mounted paintings.

Fig. 2.1a: Naturally lit corridors in the main entrance of Dakshinachitra

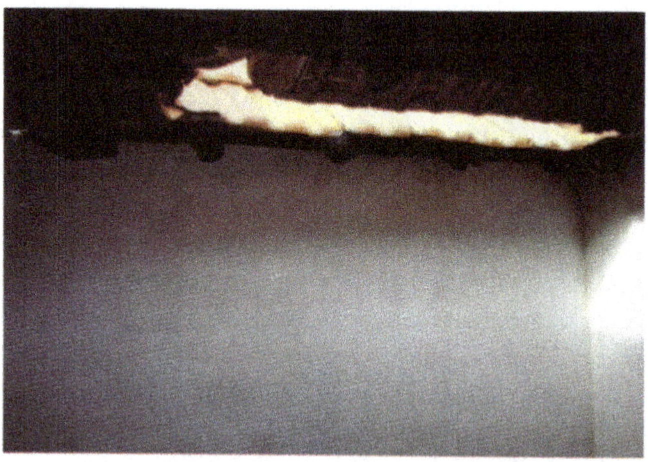

Fig. 2.1b: Vents provided for natural solar lighting in a Kerala house

In Figs. 2.1b and 2.1c we show examples from a Kerala house in Dakshinachitra where natural lighting has been used intelligently whilst maintaining the aesthetic appeal of the rooms concerned.

Fig. 2.1c: Natural lighting in a study room.

Effective daylight strategy therefore requires fenestration, plan shape, internal finishing and partition layout for optimum daylight entry and distribution and avoidance of unwanted solar gains, both of which have been brilliantly achieved in the Savitri Bhavan, in Auroville, as shown in Fig.2.1d below.

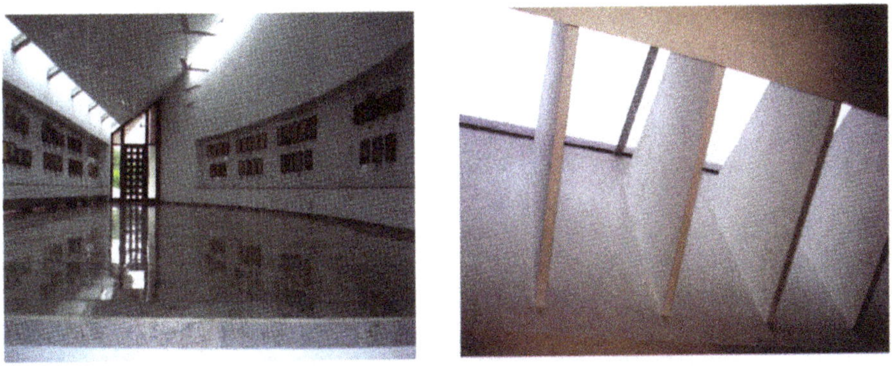

Fig. 2.1d: Efficient use of daylighting at Savitri Bhavan

The potential energy savings from daylight are large. The greatest savings in energy occur in spaces where illumination in daylight is preferable to artificial illumination.

Fig. 2.2: Efficient daylighting in an Auroville school interior

2.3.1.2 Side Lit Rooms

The majority of rooms are daylit using vertical side windows. There is a problem in illuminating rooms this way as daylight is mostly available near the window but finds difficulty in spreading to the points furthest from it. Auroville has experimented with various methods to improve the spread of daylight by using side lit rooms. The philosophy is to change the direction of light flow so that instead of pooling near the window, it is projected up to the ceiling and reflected to the rear of the room as shown in Figs. 2.3 and 2.4.

Fig. 2.3: Interior of an Auroville house optimizing on daylighting

Fig. 2.4. Another example of a side lit room in Auroville

2.3.2 Case Study: Thermal Comfort in Residential Buildings

We now proceed to discuss the concept of thermal comfort in buildings. It is obvious that this will depend largely on the climatic zone considered. In the hot sunny climes of south India keeping, buildings cool is uppermost in an architect's mind–we illustrate this theme through a case study in Dakshinachitra.

Dakshinachitra, located in Tamil Nadu is a showcase of southern India's rich architectural and cultural heritage. The township is a conglomeration of various traditional yet meticulously planned houses of Andhra Pradesh, Karnataka, Kerala and Tamil Nadu. In this study an attempt has been made to identify the traditional landscaping practices adopted to maximize the use of natural energy flows. Now we present some fact files for the four states featured in Dakshinachitra which is essential to a study of thermal comfort of buildings in Table 2.1.

Table 2.1: Factfile of the States Featured in Dakshinachitra

S.No.	State	Latitude	Longitude	Summer Temperature	Winter Temperature	Soil types	Forests
1	Andhra Pradesh	12°41′ - 22°N	77° and 84°40′E	20°C to 41° C	13°C to 30°C	Red and yellow soil	Dense tropical forests of deciduous nature
2	Kerala	10° 00′N	76° 25′E	27.5°C to 22.5°C	25°C to 20°C	Red and Yellow soil, Laterite soil	Moist deciduous forests.
3	Karnataka	15° 00′N	75° 00′E	45.6°C at Raichur	2.8°C at Bidar	Red and Yellow soil, Black soil	Sandal wood and Rose wood
4	Tamil Nadu	8°5′N-13° 35′N	76° 15′E to 80 20′E	46.6°C to 32.3°C	13°C to 3°C	Red and Yellow soil, Alluvial and Laterite soils.	Tropical evergreen, semi ever green, deciduous and temperate forests.

2.3.2.1 Landscaping in Dakshinachitra

1. **Thick Vegetation :** One of the most convenient and easy methods to bring down the temperature levels is to have a substantial amount of vegetal growth. This has been taken care of in Dakshinachitra as is shown in Fig. 2.5.

Fig. 2.5: Thick vegetal growth found in Dakshinachitra

It is a known fact that plants generate oxygen whilst taking in carbon dioxide to produce food thereby purifying the air. Thus, landscaping with trees and other plants can help minimize the bad effects of pollution. In addition to purifying the air, plants can help disperse fog, reduce wind speeds, reduce noise effects and control erosion.

2. **Water Channelling:** Water bodies (stagnant and flowing) can be seen in the whole of Dakshinachitra. Fine water droplets carried into the atmosphere by the wind affect the microclimate and subsequently bring down the temperature of the surrounding environment. We show below in Fig. 2.6 an example of a water channel that enables the flow of storm water around residential premises in Dakshinachitra houses.

Fig. 2.6: Water channelling in Dakshinachitra

Fig. 2.7: Clustered Brahmin houses in Dakshinachitra

Fig. 2.8: A typical Tamil Nadu house with an elongated form and high ceiling

2.3.2.2 Ventilation and Daylighting in Dakshinachitra

The houses shown in Figures 2.7 and 2.8 have elongated forms. Due to a higher surface area the heat loss rate will be high, so the building will be naturally cooled. These buildings receive more sunlight and hence natural day lighting will be improved.

Weather in Karnataka is generally cooler. Hence ventilation is given lesser importance and only the minimum number of openings required are provided. This is in contrast to houses in Kerala where the weather is hot and humid. In Fig. 2.9 below we show an example of roof ventilator in a Kerala house.

Fig. 2.9: Roof ventilator in a Kerala house

Fig. 2.10: Round house in coastal Andhra Pradesh to suppress wake turbulence effects

In Fig. 2.10 we show a round house often seen in fishermen's living areas of Andhra Pradesh. Since these are coastal houses prone to cyclonic conditions, they are usually temporary and require very less ventilation. The circular shape allows wind to infiltrate the house in a spiral, allowing just the required amount to enter the house.

Fig. 2.11: Passive zone created in a Tamil Nadu Merchant house

Though impractical and uneconomical in case of densely populated cities, landscaping has become an important design criterion for upcoming townships because of its long term and permanent benefits (e.g. Amby valley). Dakshinachitra gives a clear idea to make buildings energy efficient by increasing the number of passive zones, which are essentially areas that do not require artificial energy, by providing appropriate ceiling heights, and choosing the appropriate material for optimal thermal comfort. An example is provided in Figure 2.11 where the presence of a pillared corridor ensures a cool comfortable work area without the need for artificial energy.

Dakshinachitra gives an evidence of the great vision the ancestors of the southern part of India had. It is a proof of deep understanding of the necessity of a balanced relationship between man and nature. It thus exemplifies how they used nature to the full extent without exploiting it.

2.4 Temperature Profiling and Wind Speed Modulation

A region's temperature profile enables architects to plan buildings so that they can optimize the use of artificial energy. Table 2.1 shows the temperature profile for the four southern states of India. This information was used by the designers of the Dakshinachitra complex. It must be borne in mind that for cold climes it is imperative to retain building heat as far as practicable. Again a detailed study of the zone's temperature profile is essential. In addition, often calculations of the country's degree days are often needed. This is calculated by summing the number of 24 hour periods over which the external temperature falls 1°C below the base temperature (15.5°C) for a cold clime. To understand degree days better, consider an example for a cold clime. If the external temperature is 13.5°C for 12 hours, this is equivalent to 12 hours × 2°C below base line which is equal to 24 degree hours or one degree day. It follows therefore that the larger the number of degree days, the more cold the climate and the greater the demand for space heating. In cold climes therefore, optimum solar access should be given due consideration–in Fig. 2.12 we notice that although there are groups of houses, the layout is such that not even a single house is overshadowed Hall and Nicolls (2006).

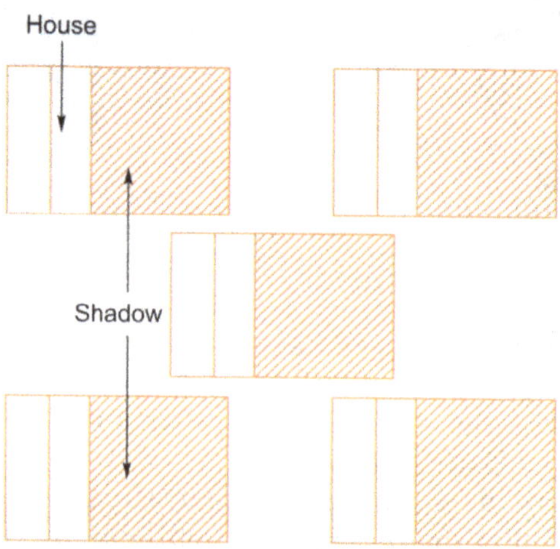

Fig. 2.12: Positioning of 5 houses for optimal solar access

Although much of India is warm, winters in northern India are cold–central heating systems are not at all common in the country. In rural and sub-urban enclaves, slight modifications in building forms can lead to comfortable indoor temperatures during the daylight hours. The first point to be considered is that any solar heating effect is swept away if winds are allowed free access to the site. One way of trapping a pool of heated air is to form single storied houses (very common in rural India) into a sun trap arrangement (Fig. 2.13). The arm lengths forming the sun trap should be carefully optimized–preferably, they should be kept short–this will reduce overshadowing effects, particularly during the early mornings and the evenings. Village houses in rural India most often have mud and clay walls–effect of solar absorption by such building materials and protection from wind-increases can easily average temperatures in the vicinity.

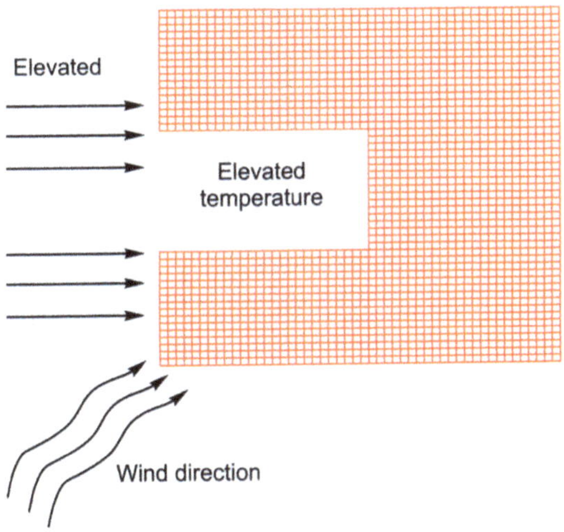

Fig. 2.13: A C-shaped building form inducing a sun trap

In a cold temperate zone, if a proposed building is slated to be sited over an area with temperatures lower than the average ambient temperature outside, it will lead to higher energy demands for space heating – this comes with a price. Some of this higher fuel cost can be mitigated by adding specially considered insulation on the building – the payback cost will then be easily recouped from the energy savings made. Additionally, having recovered the capital cost, these savings will recur for years to come.

Insulation is therefore an attractive low technology option tailor made for energy savings – hardly any monitoring or adjustments are required unlike many other electronic heating control devices. Let us now compare the heat losses across 1 m^2 of masonry wall with a U value (thermal transmittivity) of 0.35 W/m^2 K in London and Moscow. In London, 10.3 kWh of energy will be required over a heating season covering all the winter months to maintain a comfortable 21°C inside – an identical wall at Moscow would require at least 1 kWh of energy which is 14% higher than the former. The stated wall over Moscow would therefore require some additional insulation to reduce its U value to 0.31 W/m^2 to have an identical heat loss as in the London wall. This principle is equally tenable to other building parts – roofs and floors.

Heating energy use depends largely on the external temperature and therefore needs to be accounted for in order to properly monitor a building's energy performance. This is important to see whether the heating system is behaving properly and to judge and benchmark the energy use of a building normalized to some base weather. The integrated difference between the external temperature and the building base temperature where the external temperature is below the building base temperature gives the 'heating degree days' for a building.

A well run building should have an energy usage which is proportional to the number of degree days for some time period. The slope of the plot of energy use versus degree days (the performance line) has a slope which is the building energy loss rate. The performance line crosses the *y*-axis at the building base energy use (energy use which is independent of heating). The building base temperature is crucial to a rigorous analysis using degree days.

2.4.1 Wind Speed and Direction

A pressure difference is created between the windward and leeward sides of a building when it is subjected to wind flow. This effect can deliver cool or warm air into the building and can concomitantly expel warm or cold air out from the other side (Fig. 2.14). This natural flushing mechanism is used in buildings worldwide. The energy used to heat or cool the internal air is then easily lost from the building and must be avoided. With increasing wind speeds, the pressure difference across the building also increases–this leads to an immediate increase in the overall ventilation rate.

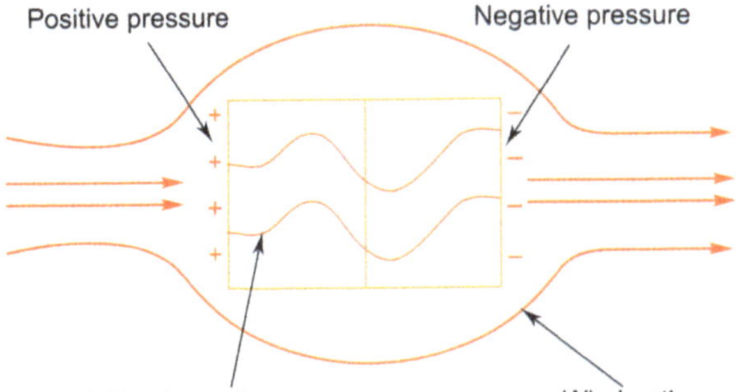

Fig. 2.14: Wind induced infiltration indicating regions of high and low pressure(adapted from Hall and Nicolls 2006)

2.5 Landscaping

2.5.1 Types of Landscaping

1. Natural landscaping, also called native gardening, is the use of plants, including trees, shrubs, groundcover, grass which are indigenous to the geographical area in which the zone is located, as well as rocks and boulders in place of groomed lawns and planned planting beds to blend residential or commercial property into the natural surroundings of the particular area. In Fig. 2.15 we show elements of natural landscaping in a Kerala house in Dakshinachitra.

Fig. 2.15: Natural landscaping in a Kerala house in Dakshinachitra

2. Artificial Landscaping of the built environment refers to the man-made surroundings that provide the setting for human activity, ranging in scale from personal shelter, to neighbourhoods, to large-scale civic surroundings. An example is shown in Figure 2.16.

Fig. 2.16: Landscaping in inner courtyard, Dakshinachitra

Necessity of Landscaping: Landscaping significantly alters the microclimate. If the microclimate is in harmony with the structure then the required alternative artificial measures for ventilation and lighting will be greatly reduced. This will reduce the cost incurred from artificial systems and lead to acceptable low levels of carbon emission.

2.5.2 Factors Affecting Landscaping and Microclimate

1. **Wind intensity and direction:** Excessive and high speed ventilation flows causes discomfort to residents. This requires effective wind channeling using natural/artificial landscaping elements. The Fig. 2.17 shows the wind profile for the Indian subcontinent. Builders and architects routinely use such information for creating houses that make use of existing wind flow patterns.

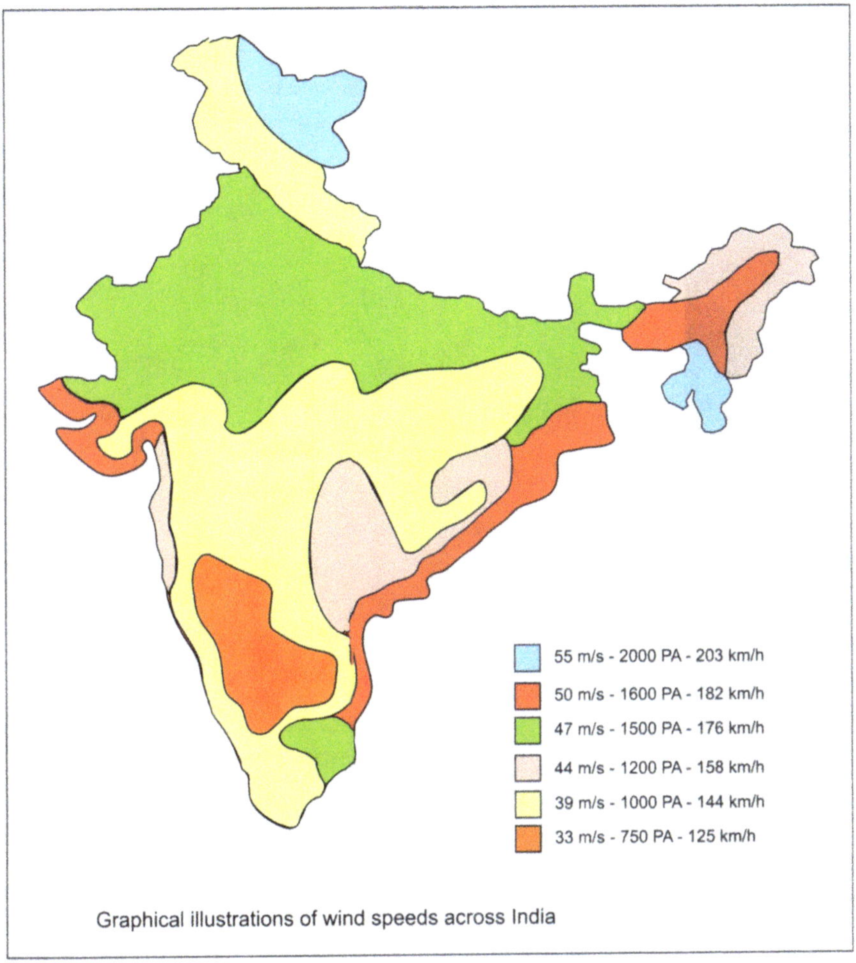

Fig. 2.17: India–Wind profile

(Image courtesy http://law.resource.org/pub/in/bis/is.nbc.2005.html)

2. **Local temperature profile:** The temperature profile and the degree days maintain an inverse relationship, *i.e.* a hotter climate is marked by lesser degree days and vice versa. Proper provisons need to be adopted while designing the landscape. One must ensure that the difference between the inner room temperature and the outer atmospheric temperature is at a minimum. This will reduce heat losses. Figure 2.20 shows the average temperature distribution over India.

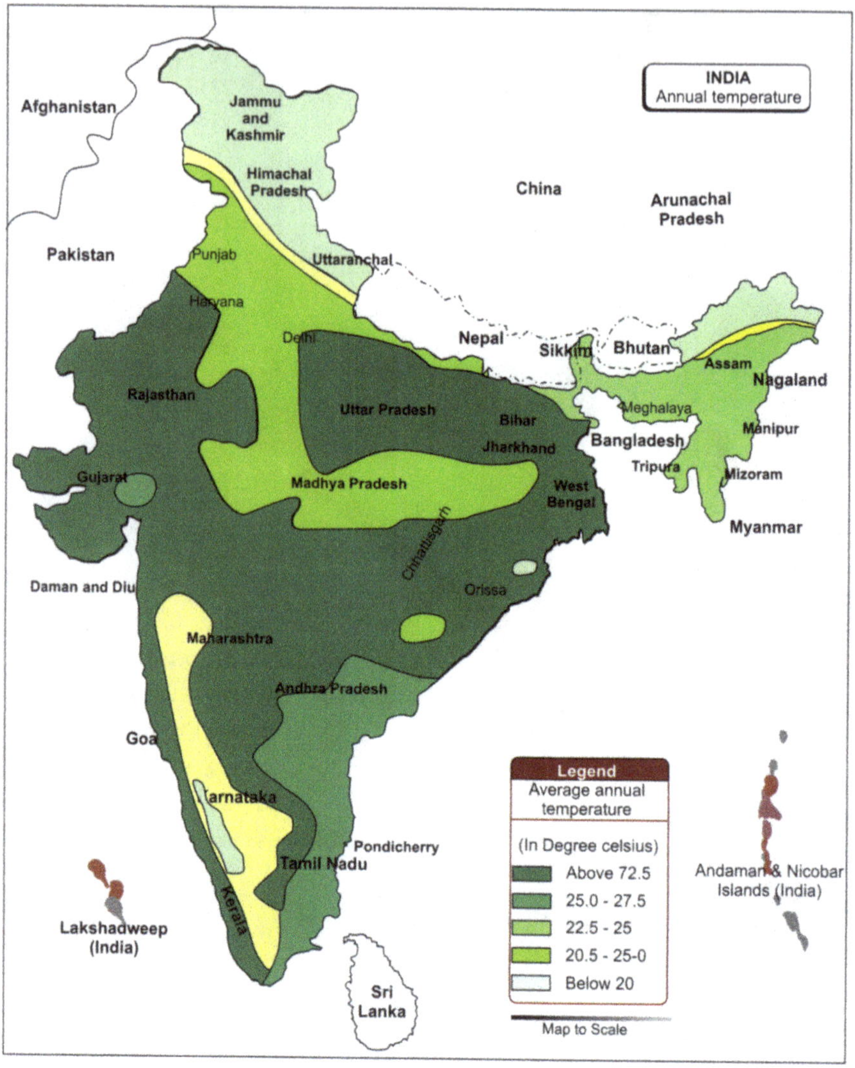

Fig. 2.18: India–Temperature Profile (www.mapsofindia.com)

3. **Land Profile:** Land slope must be taken into consideration for proper drainage. If the site consists of multiple units then overshadowing also becomes an important criterion for design.
4. **Soil Type:** Soil types can also affect landscaping and microclimate – clay laden heavy soil can function like pavements modulating temperatures close to the ground. In contrast, soils with many air pockets embedded, can trap heat beneath the top soil – This

might lead to frost at ground level and can often be undesirable. The steepest gradients in the temperature and humidity profiles occur just above and below the surface level. A soil's thermal conductivity is largely controlled by its water holding capacity which in turn depends on the type of soil. For clay, the thermal conductivity in W/mK is 0.25, 1 and 2 respectively for moisture contents of 0%, 10% and 50% by volume. Similarly, thermal conductivities vary greatly with moisture content – sand thermal conductivity values vary between 0.15 – 0.25 when the sand is dry and 0.25 – 2.0 when the sand is moist. This latter fact shows that sand has a greater propensity for heat retention as compared to clay.

5. **Radiation:** Short wave solar and long wave terrestrial radiation vary greatly, spatially and temporally, affecting indoor comfort levels. Since radiation is modulated by objects within the landscape, it is a key design parameter.
6. **Humidity:** For all practical purposes, humidity may well be assumed to remain constant over large stretches of landscapes – this happens because of efficient turbulent mixing of the boundary layer above. When humidity levels are high, evaporative energy losses from a person are considerably lower.

When humidity levels are low and air temperatures are high, evaporative cooling is most effective. Established isolation techniques can be used to heighten or lower humidity levels.

From the preceding discussions, it is clear that advantages of landscaping are manifold – it cleanses air thus increasing labour productivity, leads to energy savings, enhances the asset value of the property and above all lends a greater sense of privacy and security.

2.6 Precipitation Profiling and Driving Rain

A country's precipitation pattern has a significant bearing on sustainable architecture. Mid-latitude countries receive year-round, though milder precipitation intensities as compared to monsoon dominated countries. India's climate is affected by two seasonal winds–the northeast monsoon and the southwest monsoon. The north-east monsoon, commonly known as winter monsoon blows from land to sea, whereas south-west monsoon, known as summer monsoon blows from sea to land after crossing the Indian Ocean, the Arabian Sea, and the Bay of Bengal. The south-west monsoon brings most of the rainfall during a year in the country. Fig. 2.19 shows the average annual received rainfall over India.

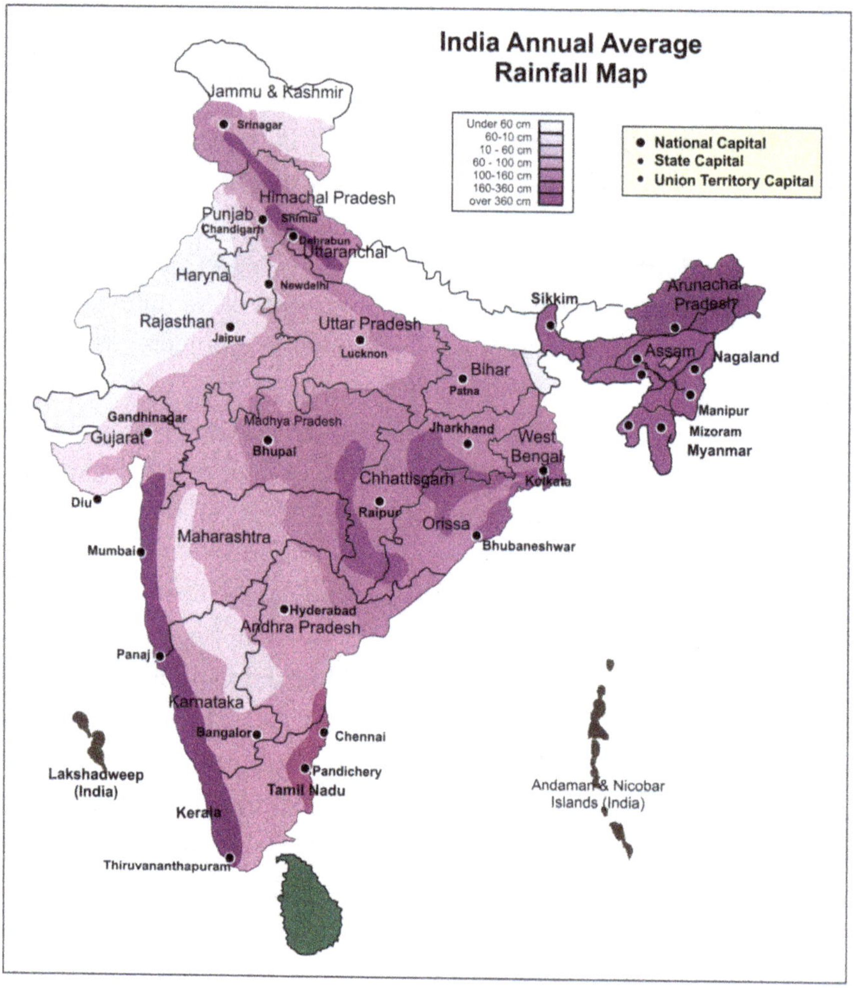

Fig. 2.19: Rainfall map of India (www.mapsofindia.com)

If one considers a vertical plane slicing through the atmosphere, and then collects all the rainwater passing through that plane, then what one is collecting is actually 'Driving Rain'. The smallest raindrops are easily swept away by the prevailing winds because of their low terminal settling speeds; the larger drops adopt a curved trajectory depending on the relative magnitudes of the terminal velocity of the drop and the strength of the crosswind (see Fig. 2.20).

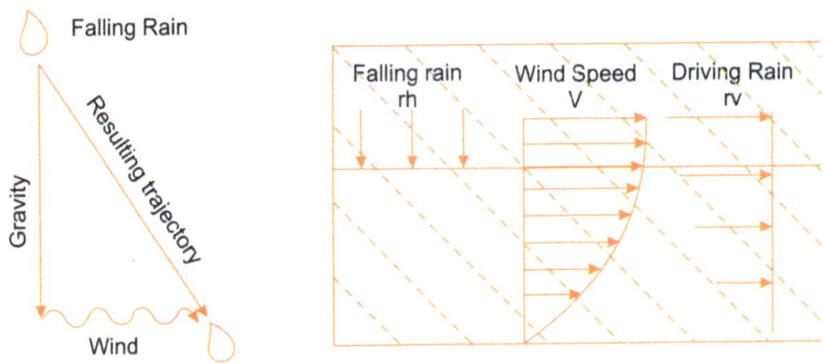

Fig. 2.20: A schematic diagram illustrating the concept of Driving Rain

By far, water deposition on buildings through rain showers overwhelms all other sources of moisture–this is true for most countries, and particularly true for India where one encounters copious monsoon showers. Rain water can be absorbed by the cladding and can lead to deterioration of the building envelope. Mould formation and staining, rotting of building fixtures are common in these situations. The above effects can be easily avoided when one understands their origins–deflection methods are used to reduce water accumulation. Slopes can also drain water away followed by drying of all wetted surfaces. In order to succeed in such ventures, architects must have a balanced knowhow of wind speeds and directions, rainfall intensities, rainfall durations and frequency, vis-à-vis building parameters including the form, height, orientation and wall design.

2.7 Solar Radiation and India's Solar Profile

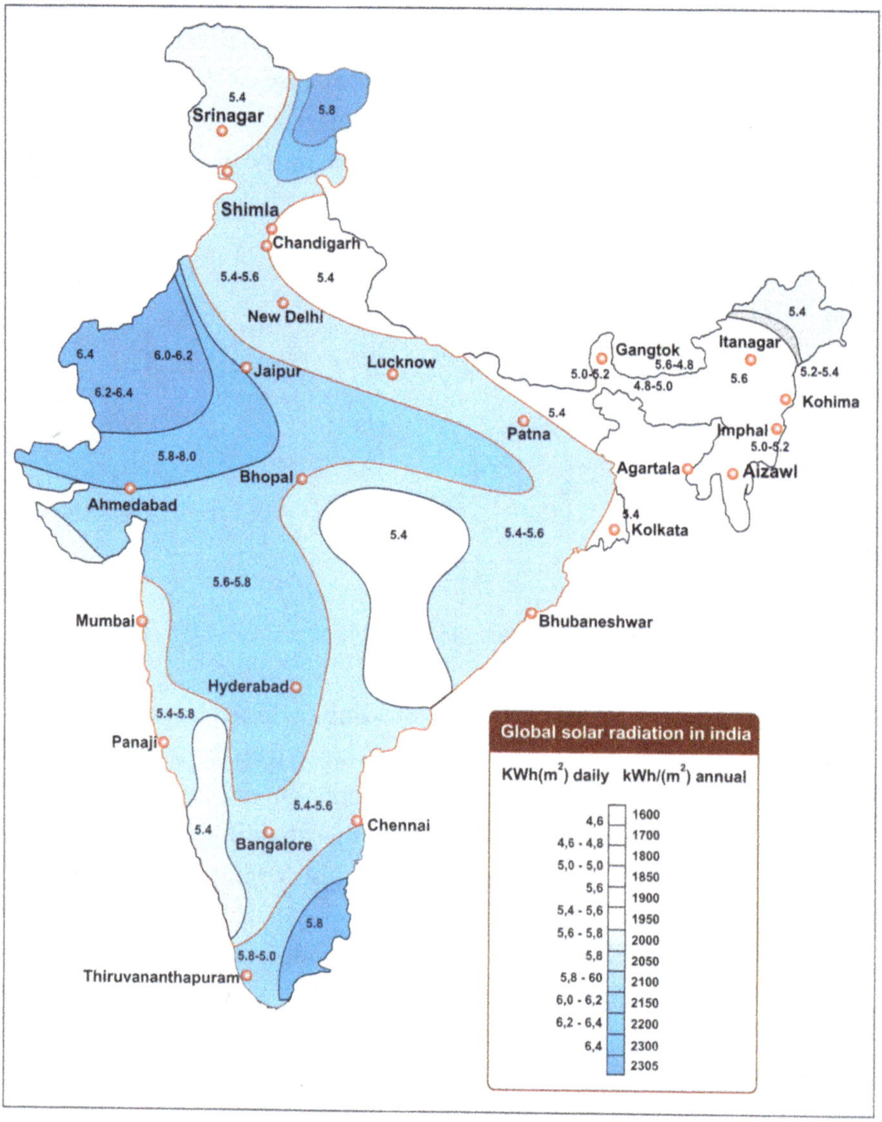

Fig. 2.21: Solar radiation in India (www.mapsofindia.com)

In Fig. 2.22 we show nuances of day lighting in a well-known house in Ahmedabad, Gujarat–The Parekh House. It is divided into functional spaces varying with time, including a zonal reversal to accommodate changing seasons. The Summer Section of the house is cooler by 2°C on the hottest day as compared to the winter section. The central section therefore is more suitable for use during the summer–flanked by two other sections preventing any direct solar exposure on the central façades. A careful play of shading and fenestration thus creates comfort zones throughout the year.

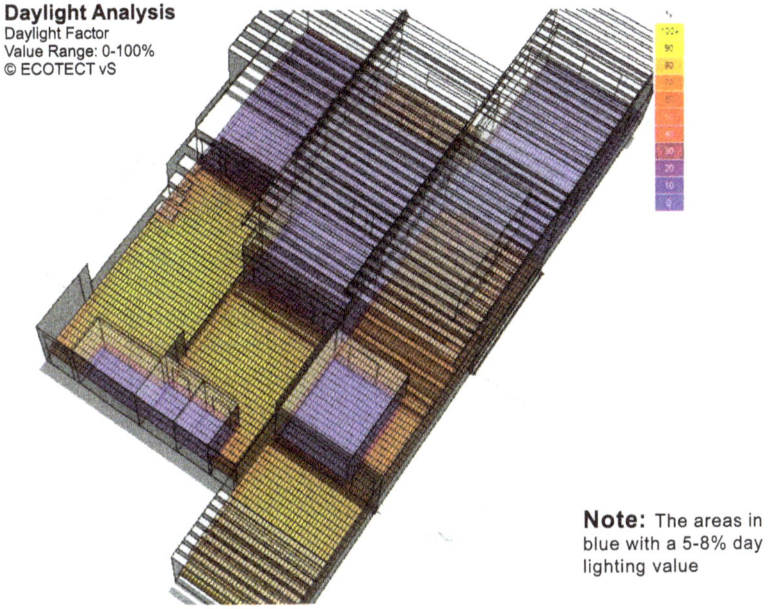

Fig. 2.22: A Daylight Analysis of the Parekh House in Ahmedabad during Peak Summer showing the Formation of Passive Zones.

Fig. 2.22. shows a day lighting analysis for a day in summer showing how well the building is naturally lit. Solar insolation has a delayed effect on the temperature of the interior spaces *(Ghosh et al 2013)*.

2.8 Infiltration and Ventilation

Ventilation is the most important factor for ensuring comfort. This is of even greater significance for dwellings in hot tropical climes, as in much of the Indian sub-continent. Ventilation can be effected even in clustered houses by providing lanes in between a pair of houses as is depicted in Fig. 2.23 showing a group of houses in Dakshinachitra. The more obvious forms of ventilation in the form of windows is well known the world over, as is the use of chimneys facilitating vertical ventilation (see Fig. 2.24 showing a house from Dakshinachitra).

Fig. 2.23: Lanes allowing free flow of air between buildings

Fig. 2.24: Chimney encouraging vertical ventilation

Fig. 2.25: A typical example of cross ventilation

In the house shown in Figure 2.25, another south Indian house from Dakshinachitra, free flow of air is achieved throughout the interiors where air is continuously refreshed. In addition, this arrangement has an added advantage–maximum day lighting is obtained through both the openings. Whilst cross ventilation is encouraged in hot and humid environments, in other climates, as in parts of Karnataka, where the nights are often chilly, the windows are preferred to be kept small to minimize cold draughts (see Fig. 2.26).

Fig. 2.26: Karnataka house with small windows to suppress the ingress of chilled winds

2.9 Building Cooling

2.9.1 Passive Cooling

Passive cooling refers to the addition of features in and around a built environment without expending extraneous sources of energy. Depression of external temperatures will also reduce cooling energy consumption. This strategy has been adopted in Auroville in south India and involves shading external spaces from the influence of the sun and the introduction of evaporative cooling. Both of these functions can be provided by trees. Fountain and water features can also reduce ambient temperatures. Figure 2.27 shows the use of water features in an Auroville guest house.

Fig. 2.27: An aesthetically pleasing and a kind microclimate for the Afsaana's Guest house in Auroville

A source of fresh air is a vital requirement for all buildings. It provides oxygen for breathing, dilutes and displaces pollutants, avoids condensation by removing moisture and keeps buildings cool. Natural movement of air through cracks, gaps, and porous elements of the building envelope is known as infiltration. The air movement is driven by pressure difference created by the wind and buoyancy of warm air. Ventilation can be either natural when it is the movement of air through purpose built openings in the fabric by naturally occurring pressure differences, or mechanical when driven by fans through ducted openings. Most Auroville buildings ensure efficient ventilation (see Fig. 2.28) and passive cooling techniques (as shown earlier in Fig. 2.27).

Fig. 2.28: An Auroville classroom using natural ventilation and lighting

References

1. http://www.eci.ox.ac.uk/research/energy/degreedays.php
2. http://www.wikipedia.org
3. http://www.mnre.gov.in
4. http://www.mtcc1170.com/images/raincontrol1170.pdf (Dr. John Straube)
5. http://commons.wikimedia.org/wiki/File:India_annual_rainfall_map_en.svg
6. http://www.buildingscience.com/documents/digests/bsd-148-wufi-simplified-driving-rain-prediction/
7. Ghosh, S., Mukherjee, S., Panambur, S., and Samaddar, A. (2013) Microclimatic and Culture-specific Adaptations of Building Forms in Gujarat: From the Indus Valley to Modernism. *The International Journal of the Constructed Environment*. Common Ground Publishing LLC, Urbana Illinois, USA. ISSN: 2154-8587
8. Hall, K. and Nicolls, R. (2006) The Green Building Bible Technical Ref. Green Building Press

ooo

3

Building Form

Chapter Overview

The shape of a building has a paramount impact on the material and energy requirements at the time of construction and even during the hours of operation. The form (shape) of the building, the land area to volume ratio, the orientation of the building, *etc*., all contribute to a determination of the quantity of material required for construction, and the energy requirements of the building when it would be operative. Structural designers and architects need to do an environmental and economic analysis of different forms and designs of buildings to arrive at the optimal shape that would cut energy losses to the minimum whilst maintaining its aesthetic credibility. Implementation of energy-efficient building technologies and efficient construction techniques trumps the conventional methods of construction over reduction in wastage of material and long term gains related to energy consumption. Reduction in energy consumption is associated with reduction in cooling, heating and lighting requirements inside the building. Therefore, the primary objective is to emphasize sustainability as an integral part of the form and construction of the building.

Factors that influence the form of the building:

Cost–buildings with straight lines are easier to construct than those with non-linear shapes.

Planning–grouped units may be required to adhere to the shape of surrounding buildings rather than inherit an independent design.

Function–the intended use of a building shall determine its form to the greatest extent. Large spaces that would equate to spaces meant for public gatherings such as banquet halls, cinema theatres, *etc.*

Altering the form of a building does have its positive effects on a building. However, decreasing the amount of energy consumed in a zone may also lead to increase in the energy consumption in other zones. The paradox maybe highlighted by this that securing renewable energy flows by increasing the surface area of fabric would be inadvertently followed by increase in fabric heat loss rates. Also, increasing the volume of a building increases ventilation and cooling energy consumption. Often a tradeoff has to be made in order to obtain an overall reduction in energy consumption. This tradeoff must be informed by the energy flows in the building. For example, since domestic buildings seek space heating/ cooling more than ventilation, smaller volumes are more desirable instead of larger volumes that would be more suited to a commercial building. This chapter describes the effect that modifying the internal and external form would have on a building's energy consumption.

3.1 Traditional and Extant Forms

3.1.1 Open Spaces and Built Form

Open spaces have to be seen in conjunction with built form in a manner which could have positive as well as negative effects. A more fluid air movement coupled with increased rate of fabric heat loss/gain would be the obvious outcome in case of open spaces. The lack of open spaces could reduce heat loss or gain. Thus they can make hot, cold or humid conditions more bearable or more unbearable.

Open spaces are an integral part of built form. The orientation and quantity of open space are major questions that should be addressed. Since any built mass would have an impact on the microclimate, more so, if it is a large one, integration of a large form that does not affect the natural climate of the surroundings would be an imperative choice. Large open spaces that would affect wind movement without posing a hindrance would help in sustaining the environment's natural form.

Open spaces are prone to heat gain during daytime and this effect is compounded if the ground is hard and the exteriors of the buildings are dark in colour. If, however, the ground is soft and green then less heat is reflected. Shading by surrounding buildings and trees can reduce heat gain to some extent. For summer shading, the building will have to be tall because of the high solar altitude. The orientation of a group of buildings may be designed, taking the sun's path during summer season into consideration. Heat loss through re-radiation is also observed with the installation of open spaces. During the day, buildings receive radiation from the sun and sky and at night this heat is re-radiated to the sky. In this case, the heat loss is compounded by a greater surface area left exposed to the sky. Heat is lost through the roof as well as the walls. This would however be the case if the buildings are not adjoined by their walls and are tightly packed.

Traditional settlements often involved a climate-dependent layout to minimize heat gains/loss. In hot-dry climatic conditions, a compact cluster of buildings was preferred. This design preference has undergone a transformation due to the fact that buildings nowadays use mechanized equipment for generating their heat or coolth. This however leads to a phenomenon which involves build-up of heat in cities and heavily populated zones. This is called the Heat-island effect. Keeping this in mind, the designer should always consider designing the layout for the ideal size and scale. A very large open space would also result in increased heat gains. A small space would have reduced heat losses at night. Other features, such as texture of the building fabric, green cover, light colours and shading maybe utilized to further enhance the thermal comfort of the occupants.

Building orientation plays a crucial role in buildings which are situated in hot climes. The orientation should be determined by taking the wind patterns in consideration. The row or cluster should inititate the funnel effect which maximizes the airflow within that zone.

The planning and designing stage is imperative to the functioning of the building and the comfort of its occupants. A building may serve a single or multiple purposes and in this aspect, its components should be multi-purpose as well. The building's fabric, thermal mass, orientation, surface area, internal space, fenestration and textures would have to be worked out for establishing clear parameters of the building's efficiency in its respective climate.

Heritage buildings such as Bharat Bhavan and Gohar Mahal in Bhopal, give proper response to climate in terms of site and building design. The buildings designed have created their own domestic microclimate to control the surrounding environment by means of certain natural techniques of passive cooling and heating, such as, incorporating in their architecture a courtyard, chhatris and shades with beautiful carvings, small windows for ventilation and also for providing scenic beauty of the nature.

3.1.2 Auroville

Auroville has experimented with building forms right from its inception. Starting from the basic 'A' Huts in the early 70's, Auroville architectural style is now famous the world over – combine unique forms built with sustainable building fabrics and the resultant buildings are perfectly adapted to the local microclimate. There are many factors that have an influence on the form of a building. Cost is a major factor here and simple cubical buildings are less expensive than complex forms. Functional requirements and aesthetics also have an important role in deciding the form. Fig. 3.1 shows the Savitri Bhavan whose form is such that filtered sunlight can stream in freely owing to its open plan architecture. Thus the picture gallery inside is lit up aesthetically with a minimum use of external energy.

Fig 3.1: Savitri Bhavan: Ranch type design with added features to optimize daylight gains

As buildings become more complex, rules of thumb of aspects such as effective distance of cross and single sided ventilation become less reliable. Large buildings are very expensive and house many occupants, so it is important that the theoretical airflows considered at the design stage are achieved in practice when the building is constructed.

Fig 3.2: A Building in Auromodele demonstrating the nuances of a superior building form

In Fig. 3.2, a functional residential dwelling is shown. Note the sweeping curves and open spaces encouraging the free flow of air. The form is such that it receives optimum daylight without heating the building excessively which minimizes the use of artificial lighting and cooling.

3.1.3 Jaisalmer

3.1.3.1 Uneven Building Form

In case of an uneven form of a building, radiative heat loss from the building to the external environment is impacted. In similar forms, the heat loss component is assumed to be minor since the adjacent or surrounding walls will "see" other walls and achieve the same temperature as them. However, they may not be applicable in all cases. This may differ in cases where the uneven form results in the radiative loss through an increased wall area. This will lead to a much cooler building as compared to one which is bunched together. A case study that exemplifies this is Jaisalmer, due to its uneven layout.

Fig 3.3: Uneven building form as observed at Jaisalmer

There is a stark difference in energy usage between residential and commercial buildings – Auroville continues to explore both these areas. It is well known that as the building size increases the surface area to volume ratio decreases. Most Auroville buildings have experimented with these ideas in such a way, that the internal space in that building although in contact with the outside world minimizes the entry of hot air from the outside, but takes advantage of solar gains, natural light and ventilation. Auroville has consistently avoided the use of deep plan buildings for residential purposes for this reason. By optimizing the above natural energy flows, Auroville architects have sought to obtain a net improvement in the energy consumption of the buildings. They have aimed to achieve the following balance:

Solar gains + reduction in artificial lighting + reduction in mechanical ventilation > Increase in fabric cooling losses

3.1.4 Gohar Mahal

Gohar Mahal was built in the 20th century and is located in Bhopal–the City of Lakes. Bhopal has a composite climate having diurnal seasonal changes between long hot dry periods to shorter periods of concentrated rainfall and high humidity in between winters with dry sunny days and uncomfortable cold nights. These climatic changes during the cold season cause the effective temperature to be much lower than in the two warmer seasons and thereby the physical comfort to depend on the prevention of heat loss from the body especially at night. In the hot season, since the heat dissipation is inadequate, the designer attempts to increase it as much as possible. Hence, it is a difficult task for a designer to build a suitable built environment. It is a general belief, that our ancestors realized the characteristics of different climates and built their buildings accordingly, using certain passive techniques of heating and cooling after fully exploiting the natural surroundings.

Fig 3.4: Gohar Mahal, Bhopal

3.1.4.1 Description

The building consists of multi-level planning, with the different levels gradually increasing in height towards the south west corner. The building has three floors with the third floor at the south west corner. There are two courtyards with a fountain in the external courtyard. It has massive walls about 80 cm to 100 cm in thickness. An interesting feature of this building is that it has road level entries at all floors. The main entry is towards the lake side in the eastern corner at ground level, and the back entry is at the second floor level in the north-west corner of the building.

The tectonics of the Building are enlisted below:

1. Evaporative cooling due to dense vegetation in the surroundings.
2. Thick walls, massive columns and heavy roof.
3. Maximum openings facing the windward direction.
4. High openings to facilitate the stack effect.
5. Wind catcher in the prevailing wind direction which ventilates the interior.
6. Conduits in double wall to circulate perfumed air in the interiors.

An array of corridors along the circumference of the rooms play the role of buffer spaces that temper heat gains in the rooms. Chajjas (Eaves) on top of window openings ensure shading and parapet walls made from bamboo provide a clear path for ventilation around the corridors. A feature known as Jharokha (an overhanging balcony) captures and redirects winds towards the interior. A light coloured exterior finish minimizes heat gains as well.

3.2 Form Dependent Natural Energy Utilization

Parameters that play a crucial role in determining the energy consumption of a building are shape and its surroundings. Careful analysis will reveal that optimization of these two characteristics would result in heat gain in cool climates and coolth in hot climates. The geometrical proportions of a building maybe such that they may have same volume but different surface areas. This results in different rate of heat gain and heat loss.

The layout of the buildings should be in a manner where each building is provided with shade from another. The type of building clusters would further determine the "quality" of shading. Martin and March (1972) have classified building clusters in three types:

1. Pavilions
2. Streets
3. Courts

Pavilions are standalone buildings either by themselves or in a cluster located in open spaces. Streets are rows of buildings aligned in a parallel fashion which are separated by a gap (which maybe an open space or a street). Courts are open spaces which are enclosed by buildings from all sides.

Buildings situated at the end of their layout or those situated in the corners of a multistoreyed buildings have a greater magnitude of exposure to outside environment due to more than one external wall. This increases their propensity for heat gain and heat loss. Therefore, such buildings/flats/rooms need to be insulated in order for them to 'perform' at the same level as other entities which do not experience high gains or losses. Insulation of building fabric will be discussed in Chapter titled 'Fabric'.

3.2.1 Buildings with Courtyards

The necessity of thermal comfort in hot climatic conditions has resulted in various building typologies such as, detached units, multi-story buildings, buildings with courtyards, rowhouses, *etc.* Buildings with internal courtyards are considered the most "appropriate" in many hot regions, especially in deserts. Many of the Dakshinachitra buildings shown in Chapter 2 have ample courtyards (Fig 2.11).

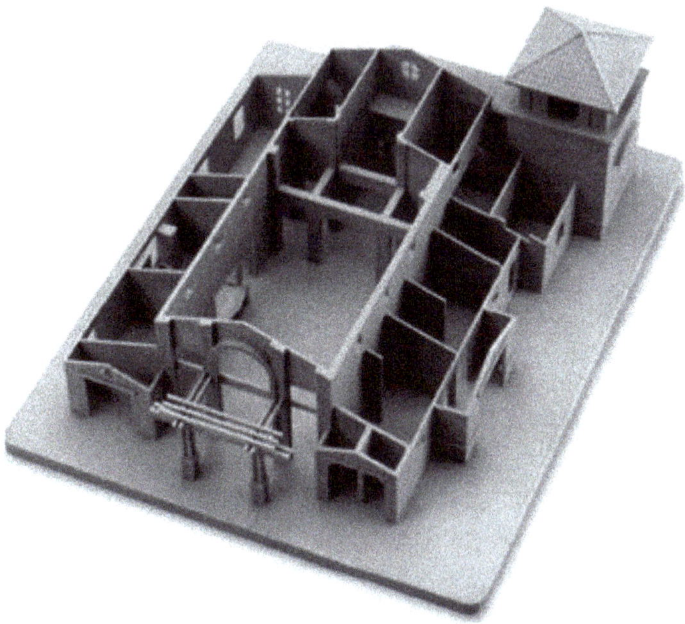

Fig. 3.5: A planned courtyard that exercises control on the placement of surrounding rooms (Source: www.javelin-tech.com).

The concept of courtyard buildings took shape due to the absence of air conditioning systems. Various cultures such as the Japanese, Chinese, Spanish and Indian adopted this concept and made it an extremely significant part of their lifestyle as well as their cultural heritage. A courtyard design not only provided relief from the harsh summer weather, but also integrated the environment with the structure. Indigenous (originating in a particular place) courtyards (Fig 3.5) ameliorate adverse climatic conditions through the nature of their placement and form the case study of (Azzawi, 1996) houses in Baghdad, states the significance of courtyards in the region and how this concept has been refined through trial and error to achieve the desirable thermal conditions.

Courtyards which are either attached to the structure or enclosed, have been attributed to have a powerful impact on the microclimate of the built form, hence being listed as microclimate modifiers. The statement however stands true only when certain conditions are met—relative dimensions of open space and built volume along with the orientation of the open space (Meir, 1995). The essence of achieving cooling in harsh environments factors in the functioning of the occupants in regard to keeping the windows closed during the day and opening them during the night, keeping the courtyard moist by sprinkling water in it, *etc*.

After flourishing as a concept and in applications, the visibility of courtyards dwindled down as compact designs start to gain favour with the masses due to its closed form. Even the traditional buildings in India underwent certain architectural changes that closed-off portions of the courtyard. Air conditioners are installed for cooling purposes which require a well-sealed environment. This change has bought with it, a certain intolerance towards harsh weather. Spaces without active conditioning are becoming a thing of the past and the rural sector is catching upto this as well owing to the increase in household incomes and affordability.

3.2.2 Solar Gain and Day Lighting

Daylight is the preferred form of illumination in buildings. It gives building occupants contact with the outside world. Weather variations and the passage of the day are revealed in the changing patterns of daylight. Of most important to low energy design is the fact that, unlike electric lighting, daylight does not require electricity to create it. However, daylight is not entirely free. It is provided through glazing systems which are thermally poor and so there is a cost in terms of additional space heating energy consumption. This energy consumption is not solely due to provision of day lighting but is also shared by the other functions of windows which are the provision of views and ventilation. Daylight can only reduce the building's energy consumption if electric lighting is turned down or switched off in response to daylight entry.

If we consider the various forms of most common building types (shapes), it is evident that if the elongated form were aligned along the east-west axis, then one of the building's large faces would face the south. The building would experience a greater solar gain than the cubical form. The amount of solar energy gained depends on the solar intensity on the vertical face, the ability with which this energy can enter this building (solar transmission) and the area of the glazing. Apart from optimum orientation of the building and exposing the site to sunlight, there is little that can be done about the effect of solar intensity. The amount of solar gain through the glazing would depend upon the factor that is transmission coefficient of glazing. So to increase solar gains the area of glazing on the south façade must be increased. It is absolutely imperative that this be achieved along with reduction of glazing in the northern face so as to keep the amount of glazing area balanced.

Shallow spaces are usually synonymous with the domestic design which is supplied with adequate amount of natural sunlight during the daytime which cuts the need for artificial lighting by a significant amount. However, if the building form was for a large volume commercial building then square floor plans would evidently have deep spaces. The deepest area which would be at maximum distance from the windows would receive the least amount of daylight and would have to rely on artificial lights to satisfy its lighting needs. An approximate standard, a depth of 2½ times the window head height is considered to be a useful daylight penetration distance.

Electricity has an adverse impact on the environment especially when it is derived from fossil fuels which have considerably high carbon emission factors. If pollution that stems from a building's energy requirement needs to be curbed and alternative sources of energy is not an option, then reduction in the energy usage is the only option. Narrow and elongated building forms shall prove to be useful in such a dilemma. Single storey buildings can afford to have deep plans since they can be supplemented with roof lighting.

Effective daylight strategy therefore requires:

1. Fenestration, plan shape, internal finishing and partition layout for optimum daylight entry and distribution.
2. Avoidance of unwanted solar gains.

Both these features have been brilliantly achieved in the Savitri Bhavan (see Fig. 2.1d) and in an Auroville classroom (see earlier Figure 2.2).

3.2.2.1 Side Lit Rooms

The majority of rooms are day lit using vertical side windows. There is a problem in illuminating rooms this way as daylight is mostly available near the window but finds difficulty in spreading to the points furthest from it. Auroville has experimented with various methods to improve the spread of daylight by using side lit rooms. The philosophy is to change the direction of light flow so instead of it pooling near the window, it is projected up to the ceiling and reflected to the rear of the room (see Fig. 2.4 in Chapter 2).

3.3 Optimizing Volume According to Heat Loss/Heat Gain

3.3.1 Impact of form on Heat Loss

By comparing the surface areas of four buildings of different shape but with the same volume and different floor arrangements, it shall be apparent how the shape and size of a building has a significant effect on its heat losses. Volume is constant for all four shapes with the exception of the surface area.

Table 3.1: Assumed Dimensions for Building Plan Shapes

Plane Shapes	Dimensions (m)			
	L	B	H	Volume (m^3)
Square	10	10	6	600
Elongated	20	5	6	600
Cross	4.47	4.47	6	600
Bungalow	20	10	3	600

Square: This form represents the smallest practical surface area for a given volume. The form finds majority of its usage in commercial establishments and aligned group housing.

Fig. 3.6: Square shaped plan

Fig. 3.7: An example of the square shaped building plan–Kunst Museum, Stuttgart, Germany (Source: www.kunstmeuseum.li)

Elongated: In this rectangular shaped plan, the width of the building is reduced and the length is extended. This results in 25% increase to the wall area as compared to the square.

Fig 3.8: Elongated shaped plan

Fig. 3.9: An example of the elongated shaped building plan–Longhouse in Devon, England (Source: www.anatomiarquitetonica.blogspot.in)

Cross: This form is highly complex and results in 34% increase in wall area.

Fig. 3.10: Cross shaped plan

Fig. 3.11: An example of the cross shaped building plan– Berlaymont Building, Brussels, Belgium (Source: www.fnnewyork.um.dk)

Bungalow: This is a form that is most commonly seen in single story buildings. This form results in 25% reduction in wall area as compared to a square. The floor and the roof surface area have almost doubled.

Fig. 3.12: Bungalow shaped plan

Fig. 3.13: An example of the bungalow shaped building plan–a bungalow in Austria
(Source: http://www10.aeccafe.com/)

The table below shows the total surface area of each shape and compares them to the square plan building shape. It can be seen from the table that the elongated, cross and bungalow plan forms have 10, 21 and 35% more surface area respectively than the square plan.

Table 3.2: Total and Individual Surface Areas for Different Building Forms

	Area (m^2)				
	Wall	Floor	Roof	Total Surface Area (m^2)	Area relative to square plan
Square	240	100	100	440	1.00
Elongated	274	105	105	484	1.10
Cross	322	105	105	532	1.21
Bungalow	181	206	206	593	1.35

Since different surfaces have different thermal transmittance values (or U values – W/(m.K), the increase in surface area does not necessarily mean there will be increase in the heat loss rate. This translates to the fact that increase in heat loss rate will be at its peak when a surface of high U-value increases in its area. In order to put this to analysis we have to weigh each element in terms of its propensity to experience heat gain/loss. This can be achieved by multiplying the surface area of the element with its respective weighting factor/U-value. The weighting factors we will use are the typical U-values of each element shown in the above table.

Table 3.3: Assigned Quantifying Factors for Individual Elements

Element	eighting factor (Typical U-values)
Walls	0.35
Roof	0.16
Windows	2
Floor	0.25

By comparing the relative weighted surface (Table 3.4) and relative surface area (Table 3.2) it would be safe to conclude that the actual heat losses are lower. The bungalow shape is seen to be the most effective in dispensing heat while acquiring an increase in its total area. This is mainly due to the increase in area of its roof, which has the largest area as well as lowest thermal transmittance.

Table 3.4: Heat Loss Factors for Various Shapes of the Buildings

Plan Shape	Total surface area (m^2)	Area relative to square
Square	440	1
Elongated	500	1.03
Cross	521	1.16
Bungalow	580	1.23

In conclusion, adopting any plan other than square (cube) would result in higher fabric heat losses. Further, these losses might have to be countered by lending a layer of insulation to the elements.

From the above example it is evident that adopting the cubical shape in building form would be the most optimal, however, there are other methods which may also help in reducing the losses in the event where the increase in surface area is inevitable. These are:-

1. Replacing the bay-window systems by a double-glazing system
2. Adding an adjoining external buffer, in the form of a conservatory or a garage
3. Installing insulation to ceilings
4. Prevent any external features such as projections that would warp the efficiency of the square-shaped plan

3.3.2 Surface Area to Volume Ratio

Energy is utilized in a different manner by residential and commercial buildings. Heating and cooling is given more weightage in residential buildings as opposed to commercial buildings in which the primary areas of focus are generally lighting, ventilation and air conditioning due to the number of occupants and the purpose of the building. Also, the surface area to volume ratio plays a significant part in assimilation of these needs.

	1-mm cube	2-mm cube	4-mm cube
Surface area	6 sides × 1^2 = 6 mm^2	6 sides × 2^2 = 24 mm^2	6 sides × 4^2 = 96 mm^2
Volume	1^3 = 1 mm^3	2^3 = 8 mm^3	4^3 = 64 mm^3
Surface area-to-volume ratio	6/1	3/1	1.5/1

Fig. 3.14: Surface area to volume ratio decreases withincrease in the cube's dimensions

If the smallest cube was assumed to be a domestic scale building and the largest being a commercial scale building, it can be derived from the illustration that the surface area to volume ratio for the domestic scale building is 6:1 which decreases to 1.5:1 for the commercial scale building. This translates into the fact that the domestic scale building would be more in contact with its external surroundings than the commercial scale building. The drawback would be that the spaces situated on the outer periphery will experience fabric heat loss but will make up by taking advantage of the solar gains, natural light and ventilation.

In large buildings, deep spaces are fairly isolated from the outside environment due to their positioning. The innermost areas of these deep spaces will experience less heat loss but will also be inadequately supplied with natural sunlight and ventilation. This will put buildings with deep plans at a disadvantage.

3.3.3 The Effect of Volume on Ventilation Loss

The equation for steady state ventilation heat loss rate affirms that ventilation heat losses are directly proportional to the volume of the building. Therefore, reducing the volume of heated spaces would reduce the ventilation heat loss rate from the building.

$$Q_v = N.V.S_p.\Delta T$$

Eqn. 3.1 Equation for ventilation (Q_v) heat loss rate

Where,

Q_v = Heat Loss Rate

N = Number of air changes per hour; 1 air change = 1 room volume

V = Volume of room (m^3)

S_p = Specific heat factor for air = 0.34

ΔT = Temperature difference between outside and inside (°C)

Another issue which needs to be taken care of is stratification. Warm air ascending towards the ceiling creates a temperature gradient that increases as we move up towards the ceiling. The temperature at a lower level (0.0 - +2.0m) may approximately be 18°-21°C which may increase to 28°-30°C (+2.0 - +4.0m). This leads to additional energy consumption in two ways. Firstly, energy gets used to heat up unoccupied space (+2.0 - +4.0m). Secondly, the temperature difference across the roof structure increases. This in turn increases the rate of heat loss through the building fabric.

Certain measures need to be adopted to counter these issues. First one would be to reduce the ceiling height. This allows for the warm air to stay in occupied space. If structural changes are not feasible, another option that could be adopted is de-stratification fans. These fans are placed on ceiling level and tend to push the warm air downwards, when the temperature crosses a certain level in that zone.

It should however be noted that room height is form dependent and may vary according to the placement of other elements as well, such as windows, finishes, false ceiling, *etc*.

3.3.4 Relationship Between Volume and Cooling

It is evident by now that increasing the volume of a building leads to increase in ventilation heat loss rates. However, increasing the volume of rooms that are more susceptible to overheating, can reduce temperatures in the occupied zone and if such a strategy is executed in a planned manner, it would reduce the load on energy required for cooling that space.

The cause of rise in the internal air temperature is due to sources other than the heating system. These sources are the sun, lighting, people and appliances. Heat derived from these sources falls under the category of internal gains.

The Jaisalmer house not only has simple courtyards and very narrow vertical ducts, they also have staircase shafts that efficiently deflects the wind into the house. Bearing in mind that stone walls are used, it is easy to perceive that with this high thermal inertia, the addition of shafts modulates the entering airflow into the living space. Such ventilation systems lend substantial air flow through the house. Jaisalmer buildings are greatly affected by wind-blown dust and sand. This problem is partially circumvented by adding a 6m high wall around Jaisalmer. Paved surfaces, in addition, suppress the lifting and saltation of wind-blown dust. During winter, ventilation needs are at its minimum – windows are left open during the day and are tightly shut during the night. Solar heat gains in such buildings are mediated more through the roof than through the walls – the roof constructions are massive often with a thermal time lag of the order of 24 hours. The peak indoor temperatures occur over the day rather than at night when people are awake – this is the desirable feature. Jaisalmer houses generally have a basement – these acting as a secure place for storing valuables. These underground rooms have very low temperature swings inside throughout the year because of the absence of heat loads.

3.4 Ventilation and form

The inflow of fresh air is of paramount importance to a building. Aside from providing fresh air for breathing, it keeps the building cool, displaces odour and pollutants and prevents growth of mould due to sustained moisture. The movement of air through gaps and spaces in the building envelope is known as infiltration. This flow of air is governed by the pressure difference that is created by the flow of the wind itself. Ventilation maybe either natural (resultant of natural wind flow around a building) or artificial (through fans and air conditioning systems).

Natural ventilation is considered to be effective for a certain depth, depending upon the placement of ventilation points. Single-sided ventilation would be effective for a depth of 5-6m from the ventilation point. However, cross-ventilation is effective over a distance of 13m between the ventilation points. To reduce the energy consumption by artificial ventilation systems, cross and elongated plans maybe adopted that reduce the overall depth and encourage natural ventilation.

3.4.1 Airflow Around a Single Building

When moving air impinges on a building in its path, portions of it undergo changes in its velocity. The reduction in velocity causes the kinetic energy of the flow to convert to potential energy which reinstates itself as positive pressure. The magnitude of area which the positive pressure shall envelope, depends upon the shape of the building. A streamlined building shall develop a small area of positive pressure. However, in case of a building with a large face that affronts the moving air, the region of positive pressure would be comparable in size to that face.

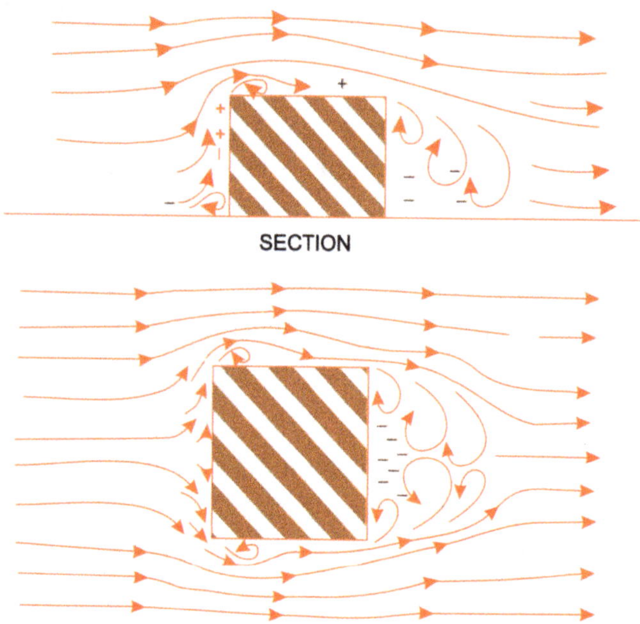

Fig. 3.15: Airflow around a single building

Incoming air is pressed around, above or below the building resulting in an increase in its velocity whilst the potential energy due to the positive pressure gets converted to kinetic energy. As the velocity starts to exceed that of the incoming flow, the negative pressure (suctions) builds up in those particular zones.

Wind tends to hug the surface when it flows above it and when it approaches a sharp corner, it tends to turn with it. However, due to its own momentum, it is unable to, and this leads to the wind separating from the building, which is defined as the upstream limit of the wake. The wake consists of negative air pressure and relatively less air movement though the boundary between the wake and the free stream has significant amount of turbulence in it. The transfer of momentum across the wake boundary blurs its position. Free-stream airflows arch themselves towards the wake from all sides at which point, they rejoin the opposite streamline which is downstream of the building. This point at which the free-stream airflows rejoin is the end of wake cavity.

Negative pressure through the entirety of the wake can only be ensured if the free-stream airflows draw back and join at the downstream level. The increase in suction is directly proportionate to the speed at which the airflow draws together. The level of suction is highest where the radius of curvature of the wake boundary is the smallest. As the level of wake suction approaches zero (towards ambient air pressure), the radius of curvature of the wake cavity approaches infinity. This essentially means the structures in the wake will have no direct impact on the shape of it.

The geometry of the wake sets the outline of the air movements. In the exterior part of the wake, air movement is in conjunction with the free-stream, however, the zone within the wake maybe considered to contain a cavity of relatively still air. The pressure differences essential for initiating ventilation within the building are unlikely to occur in this zone.

Airflow around a building or a group is a fairly complex concept. Fig 3.15 depicts the prevalent pattern that is found in the case of strong winds encountering a building. These airflow patterns are beneficial in determining the ventilation on the inside of that building along with the favourable properties of the exterior fabric. CFD (Computation Fluid Dynamics) techniques have been used to simulate air flow around multiple buildings (Baskaran and Kashef, 1996) with positive results.

3.4.2 Downwash

Some of the strongest winds around buildings are found at the windward side and edges of tall buildings protruding above the surrounding general level of development. This effect occurs because winds aloft are stronger than at ground level, causing higher pressures at the top of a building's windward face than at its base. This pressure difference creates a strong downward flow on the windward face.

3.4.3 Corner Effect

Strong winds occur at building corners as the airflows from the higher pressure zone on a building's windward side to the low pressure zone on the leeward side. Accelerated wind is generally restricted to an area with radius no longer than the building's width. The taller and wider the building, the more intense the effect. If two towers of 30 stories or more are placed less than two building widths apart, acceleration will fill the entire space between them.

When a multi-storied building has an open passageway or sits atop a set of columns, air forces its way through this opening and intensifies in its flow downwind. This is known as Gap Effect.

3.4.4 Channel Effect

To alleviate the commitment of the wind, a layout of tightly grouped buildings lined alongside the streets or in an open space maybe installed. This would initiate the channelling effect.

3.4.5 Venturi Effect

When two large buildings are angular to each other, they tend to create a funnel which has a small opening and is lesser than 2-3 times the building height. The small opening accelerates the wind which flows through it to much higher speeds. The occurrence of this is termed as Venturi effect. The buildings should be at least 5 stories high with relatively large open spaces around them.

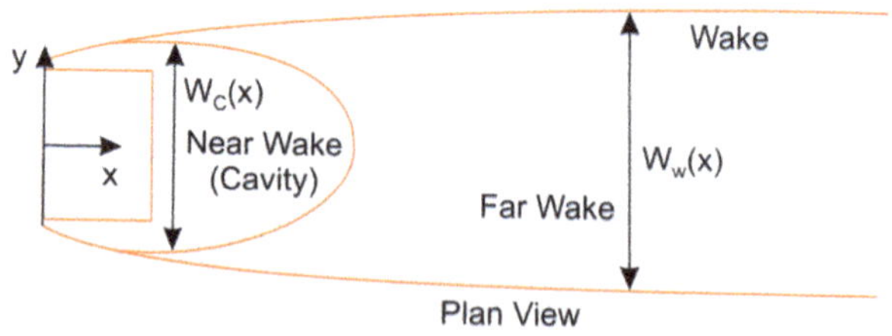

Fig 3.16: Far and near wake geometry of a building

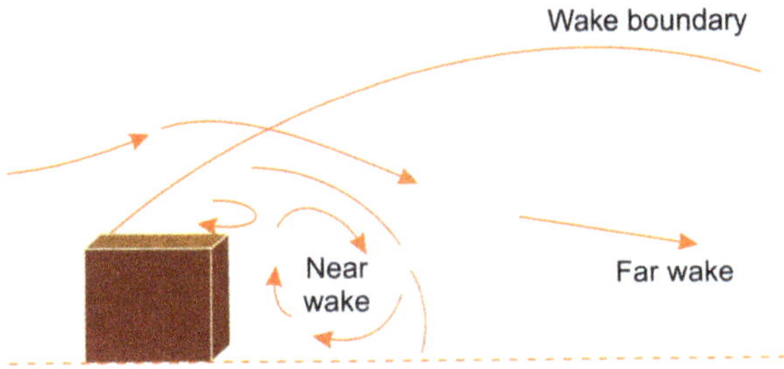

Fig 3.17: Figure depicting the wake boundary coupled with the downwash

3.4.6 Pyramid Effect

Wind faces little resistance when it encounters structures which are pyramidal in shape. The structure causes the wind to disperse in different directions. The application of this is seen in tall buildings which utilize the pyramid principle and reduce airflow wake and corner effects.

The direct relationship between form factor and airflow patterns is irrefutable. Some measures that might establish an effective balance between these two parameters are stated below:

1. A higher elevation tends to result in increased airflow between floors
2. Upper areas need to be provided with sufficient ventilation to provide an egress for hot air
3. High ceiling concept maybe utilized for its effectiveness
4. Multi-partitions need to be adopted for their versatility
5. Elements such as screens, louvers, jalousies and grills may be used in place of solid partitions
6. Properly placed windows and winged walls encourage air movement

3.5 Lighting and Thermal Method of Design

3.5.1 Loss Versus Gains–Energy Balance

To be successful in terms of utilizing available natural energy flows we need to achieve energy improvements in the following ways to outweigh any increase in fabric heat losses or use of cooling plant:

1. Use of solar gains for heating
2. Reductions in artificial lighting
3. Reductions in mechanical ventilation.

Size of the building leads to complexities in energy flows which make it more difficult to ensure if the required balance has been achieved or not. Computer modeling and simulation techniques may be used extensively here but the disadvantage is that it is time consuming to set up the model in the first instance. In such a scenario, lighting and thermal (LT) method of design maybe adopted. This is a simple method which uses graphical interface and techniques.

3.5.2 Designing Floor Layouts Using the LT Method

Comparing energy consumption between two alternative floor plans is made easier by the Lighting and Thermal (LT) method of design. For an overall reduction is energy consumption in commercial spaces, the design needs to be executed for better use of daylight, passive solar gains and natural ventilation.

As an example we shall take into consideration two distinct floor plans with a floor area of 625 m^2 in a typical office building. One of the plan is square-shaped and the other is L-shaped.

The following assumptions have been made:

1. The facade has a glazing ratio of 50%
2. Transmission factor's value can be taken as 0.35. This is usually the value adopted for buildings which are heavily shaded.

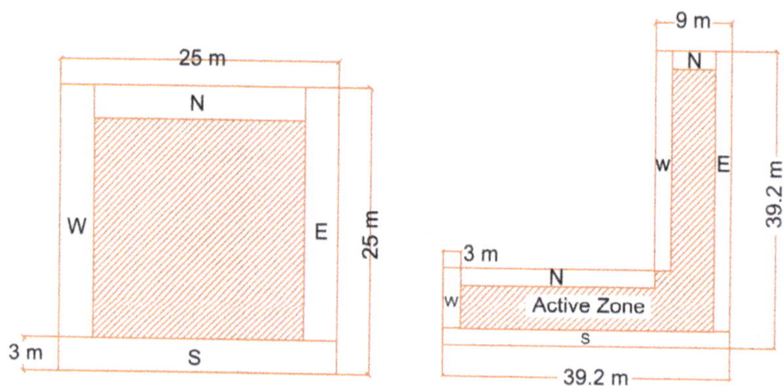

Fig 3.18: Illustration for square floor plan and L-shaped floor plan

The building is segmented into two zones: active and passive. Passive zones receive passive gains such as natural light, solar gains and natural ventilation. The depth of passive zone has been taken as 3 meters.

At the onset, we compare the areas of active and passive zones of the building as shown in Table 3.5. The square plan has 81% more active zone area than the L-shaped plan. Active zones would rely heavily on mechanical equipment to maintain thermal comfort, therefore, a square-shaped plan will endure a higher degree of energy consumption to provide the same degree of thermal comfort as the L-shaped plan.

Table 3.5: Areas of Active and Passive Zones for Two Floor Plans

Plan Shape	Area of passive zones (m^2)	Area of active zones (m^2)
Square	425.8	264
L-shape	199.2	361

An analysis using the LT method is required to quantify the energy consumption at this stage. Using the LT curves, basic mathematic calculation reveals the total energy consumption for both zones as stated in the table (Table 3.6).

Table 3.6: Predicted Annual Energy Consumption for Two Plan Shapes

Plan shape	Zone energy consumption (MWh)		
	Passive	Active	Total
Square	44.71	27.72	72.43
L-shape	20.92	37.91	58.82

Table 3.7 depicts the total energy consumption for each plan shape. Considerable reductions in lighting and heating loads are observed for the L-shaped building. It would have been safe to assume that the building with a narrow plan would have higher space heating requirement but this gets countered by additional solar gains from the south face of the building.

Table 3.7: Breakdown of Energy Consumption Predicted by LT Method of Design

	Energy consumption (MWh)	
	Square	L-shape
Lighting	6.6	9.02
Heating	2.64	3.61
Vent and cool	18.48	25.27
Total	27.72	37.91

The use of photoelectric lighting which turns off in the presence of natural sunlight translates to the largest reduction in energy consumption.

It should be noted that the technique utilizes simplified assumptions, which should serve as guidelines at the design stage. Further detailed analysis would be required for accurate values.

3.5.3 Conclusion

Cubical (square) shaped building form is favourable for residential and small commercial spaces due to space utilization. These however, should be supplemented with space heating/cooling for energy demands. Large commercial buildings should adopt narrow plans to rely comparatively lesser on mechanized equipment to reduce their footprint on energy consumption.

The LT method of design utilizes the set of curves given below (Fig 3.22). These curves help in determining the annual energy consumption for a given floor area. The energy consumption is relayed in MWh/m^2. This is further segregated into lighting, heating and ventilation/cooling energy usage. The graphs are available for four different orientations of building form as well as two different climate types (warm and cool). The one for a south facing building forming a cool climate zone is given below.

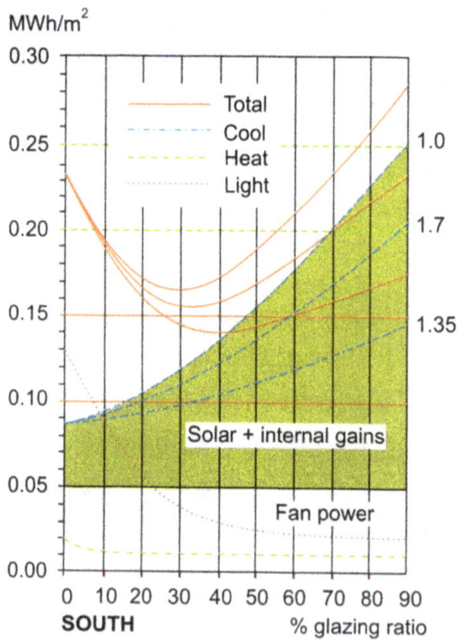

Fig. 3.19: Illustration for LT Curve for a South Facing Office's Passive Zone

The following are the features of the graph:

Heating (dashed line): With an increase in glazing ratio, solar gains increase as well leading to a decreased demand for space heating. The passive zone becomes saturated with the solar gains, thereby making any further increase in glazing ratio, redundant.

Cooling (dot.dash line): The topmost curve represents no overshadowing, the middle one, moderate overshadowing and the bottom one, heavy overshadowing. The transmission factors for the above are 1.0, 0.7 and 0.35 respectively.

Lighting (dotted line): Lighting responds to the glazing ratio and in this case, as the glazing ratio increases the requirement for artificial lighting dwindles. After a certain point, daylight saturates the passive zone and no further reduction in artificial lighting is observed.

Total Energy (solid line): The total energy line is represented by the solid line and are segmented into three curves according to the three transmission factors. Notice the U-shape of the curve which translates to an initial reduction in energy consumption owing to natural gains in solar and lighting. After a certain duration, these level out, leading to a higher consumption in cooling energy, which leads to the curve rising once again.

3.6 Proximate Alterations and Passive Cooling

Building shape and orientation are the primary elements that play a crucial role in the design of a building with natural ventilation situated in hot and humid clime. The orientation of the building would determine the intensity of solar heat radiation that is absorbed from the external facade of the building. The shape would be responsible for the exterior surface area to volume ratio and the flow of ventilation air. Both these factors work in conjunction to streamline a design's efficacy and to improve the thermal comfort of its occupants.

Fig 3.20: Eaves and Jharokhas adorning the exterior wall at Jaisalmer

Fig 3.21: Narrow alleyways that channelize airflow at Jaisalmer

In Warm/Hot and Wet (Humid) climates, shape and orientation are integral factors and must be considered together in the design process. Both shape and orientation have thermal and ventilative consequences as follows:

3.6.1 Thermal Considerations

Elongation of walls along the north and south faces minimizes solar gains and exposures to east and west should be minimized. Shading measure maybe adopted where necessary but would prove more effective if they were aligned along the east-west axis as well.

3.6.2 Ventilative Considerations

For optimum thermal comfort by cooling the occupant's body, an above average velocity of air movement needs to established in the occupied parts of the room. In case of elongated buildings, pressure difference develop if the building is perpendicular to the movement of air. To exacerbate this phenomenon, the plans of the building should be deep, in order to cool the maximum internal area.

3.6.3 Window Placement

To achieve the ideal amount of ventilation within a building, the windows are to be placed on adjacent walls, with the longer face of the building situated in a perpendicular manner to the wind. An eccentricity of 20-30 degrees will not hamper the ventilation as well. When it exceeds the above stated angles, the velocity of the interior air flow is much lower in comparison to that of velocities that originate when the building is perpendicular to the air flow.

However, a 45-degree approach angle on the wind provides maximum interior air velocity, when the windows are placed on opposite walls. This is much more efficient than in the case of wind approaching at a 90 degree angle. In case the wind is parallel to the ventilation face, it becomes fairly complex and uncertain to determine interior air velocities since it depends entirely on the wind fluctuations.

3.6.4 Elevated Buildings

Those buildings that are erected on stilts (columns) or shear (lateral) walls experience up to 30% gain in ventilated air as compared to those which are resting on the ground. This is due to the fact that wind velocities are higher as the building height increases.

This design element encourages the flow of cool, shaded air from the bottom of the building. It is an extremely popular concept in hot and humid climates.

Situating a water body next to the structure would be even more beneficial as the air cools even further due to passing over water.

When situated next to water, elevated buildings can allow cooler air that has passed over the body of water to enter the building from below. Elevating the building may also be worthwhile if the ground is continually damp or when the building is located in a flood plain.

3.6.5 Spacing

Sufficient spacing between buildings needs to be provided in order to provide a clear passage to the winds which would provide ventilation. The layout should be such that buildings do not get situated in the wake of surrounding buildings or the landscape. The built-up environment should be open as well as conducive for channeling wind flows towards the building. In case of a sloping landscape, the crest of the hill on the windward side would be favourable for siting a building. The bottom of valleys encounter extremely reduced air flow, therefore, they should be avoided as far as possible.

Street layouts can be used to channel airflow in higher density site planning. If buildings are grouped, airflow principles should be used to determine the most suitable arrangement.

3.6.6 Resolving Conflicts Between Thermal and Wind Orientations

In cases where a building's orientation has to choose between solar gains and wind gains, solar gains usually take precedence. Though, a well insulated building with a light facade and appropriate shading devices would not experience significantly lower temperatures when oriented for optimum solar gains. In such cases, the internal cooling would be further achieved by taking wind orientation into consideration.

However, it is much easier to accommodate design concepts that allow for natural ventilation in less than optimal conditions as compared to providing artificial solar control devices. The placement of windows and detailing of the facade can be designed in such a way that encourages interior air flow.

Pre-emptive measures to counter heat gain through interior and exterior points of focus.

Building Characteristics:

1. Minimize radiation effects with plans of up to 1:3 on east-west axis.
2. Create plans with one-room depth to maximize cross-ventilation.
3. Connect interior spaces with adjacent exterior patio spaces.
4. Minimize solid interior partitions.

Shade (non-landscaping) for minimizing heat gain:

1. Maximize shading of the entire building–consider using a double (fly) roof arrangement.
2. Include generous overhangs to avoid sun and rain.
3. Provide screened, shaded outdoor living areas.
4. Provide shading devices for south-facing windows, if any, to minimize solar gain.
5. Maximize north facing windows to increase daylighting without solar heat gain.
6. Install shading devices on the outside of windows rather than inside.

References

1. "The LT Method version 1.2 - An Energy Design Tool for Non-Domestic Buildings" - Commission of the European Communities - Directorate - General XII for Science, Research and Development

2. Subhi Al-Azzawi (1994, August) "Traditional neighbourhoods: A Detailed Checklist for Identifying, Analysing and Appraising the Concepts of their Passive Solar Design Characteristics regions of the hot-dry climates".
3. Elements of Intelligent Design - Thermal Comfort. Retrieved from http://http://www.archinology.com/
4. Open Spaces and Built Form, Building Configuration Retrieved from http://http://www.new-learn.info/
5. Hall, K. and Nicolls, R. (2006) The Green Building Bible Technical Ref. Green Building Press
6. Manoj Kumar Singh, Sadhan Mahapatra, S.K. Atreya (2011, September) "Solar passive features in vernacular architecture of North-East India"
7. M.B. Gadi (2010) "Application of design and passive technologies for thermal comfort in buildings in hot and tropical climates"

4

Building Fabric

Chapter Overview

4.1 Introduction to Building Fabric

Buildings lend a safe, efficient and a comfortable environment for humans to complete their tasks. The outermost layer of a building forms a barrier between the harsh exterior climes to a thermally suitable interior environment. The building fabric reaffirms its significance usually in a scenario that has seasonal extremes. Building fabric (Outermost layer) is also sometimes referred to as the building envelope. Its design is bilateral–to prevent heat loss in cold climates and to cut off heat gain and prevent loss of coolth in hot climate. A cold climate demands a well insulated fabric to retain heat whereas a thin fabric suits a hotter environment. Such a fabric plays a crucial constituent of a green building. In this chapter, we will discuss the various building fabrics, their individual performance against reducing heat losses through various elements of a building and suitable methods to enhance a building against such losses.

A building loses its thermal balance through either heat loss or heat gain depending upon the climatic conditions. There are two types of heat losses: fabric heat losses and ventilation head losses. Fabric heat losses maybe further classified into:

1. Conduction
2. Convection
3. Radiation

Ventilation heat losses are generally mediated by infiltration through gaps in fabric and joints. In order for the building to maintain its thermal state and not be influenced by the external weather conditions, 'heating' and 'cooling' systems are installed, which function at the specified level in order to maintain a specific range in temperature. A building fabric ensures that this system takes the least amount of load to maintain that state and ultimately brings down the dependency on resources such as power, natural gas, *etc*.

In order to reduce fabric heat losses from a building, it needs to be well insulated with a material that has a low value of thermal transmittance (U-value). A combination of such materials that would ultimately surround the building would enhance the performance, in terms of lowering heat losses. A building fabric may be used externally or internally depending upon the type of fabric. An external fabric may find its application in:

1. Structure
2. Roof

3. Wall Fabric
4. External Windows and Doors
5. Sunscreens and External Attachments.

Similarly an internal application would pertain to:

1. Internal Walls
2. Internal Doors and Windows.
3. Internal Joints and Finishings.

An internal application of building fabric would just be a control element, whereas the external fabric would be truly responsible for a building's overall performance. Application of glass or glazing is extremely common in commercial buildings and has now found extensive usage in residential projects as well. Glass, being a very poorly insulated element has its advantages also in terms of views, passive solar gains, means of escape and daylight. In order to reduce losses through glazing, one of the following may be adopted:

1. Double glazing (or even triple glazing depending upon orientation of the building)
2. Low emissivity glass (Low-e)
3. Integrated Blinds or Louvres

4.2 Thermal Properties of Building Materials

Heat moves from a warm place to a cold place through three different mechanisms: conduction, convection and radiation. A detailed definition is given below:

Conduction: The heat transfer process that takes place through a temperature gradient between the molecules which form the solid parts of a building fabric is known as conduction. Heat flowing in the direction of the coolest molecule can be thought of as conduction. Dense materials, where molecules are more closely packed, are found to be good conductors as compared to materials which are less dense and whose molecules are widely spaced. Lightweight materials may be bad conductors but are good insulators which find their use in controlling heat losses. A building would generally use a combination of both materials to enhance its performance in terms of structural stability and thermal comfort of its occupants.

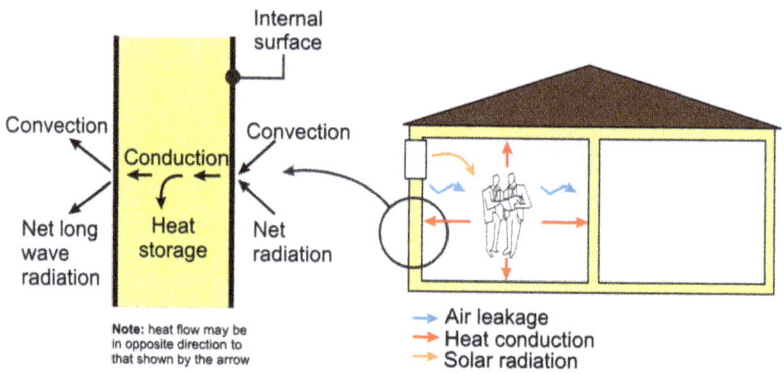

Fig. 4.1: For heat transfer mechanisms through an insulated cavity wall

Convection: Heat transfer through a fluid (liquids, gases, *etc.*) is known as convection. Air surrounding a surface that is hotter than itself would also experience a rise in temperature. Cold air is dense, therefore, it settles down but warmer air rises up. The warm air may come in contact with a surface which is cooler than itself and in that case, the heat will be transferred from the air to the surface. This mechanism is the dominating factor in glazing systems and walls.

Radiation: This is the heat exchange that takes place between a cold body and a hot one. Temperature of the surfaces and emissivity/absorptivity of the source/receiving surfaces are the primary factors that radiation depends upon. The emissivity (E) of a material (usually written ε or e) is the relative ability of its surface to emit energy by radiation. It is the ratio of energy radiated by a particular material to energy radiated by a black body at the same temperature. A true black body would have an $\varepsilon = 1$ while any real object would have $\varepsilon < 1$. Emissivity is a dimensionless quantity. Absorptivity (A) is completely the opposite of emissivity which involves the ability of an object to absorb thermal radiation.

When the temperature of different surfaces within a room increases, it is usually due to radiation between the objects such as internal walls, furniture *etc*, and the inner surface of the external wall. It also causes the heat transfer between cavities and loss of heat from external surfaces. In order to reduce these radiative transfers, emissivity of the surface is reduced, for example, by using foil backed insulation boards and low emissivity coating on the inner pane of double glazed units.

4.2.1 Thermal Transmittance (U-value)

The U-value (or U-factor), also known as the overall heat transfer coefficient, states how well a building element conducts heat. It measures the rate of heat transfer through a building element over a given area, under standardized conditions. The usual standard of temperature gradient is considered to be 24°C, at 50% humidity with no wind (a smaller U-value is better).

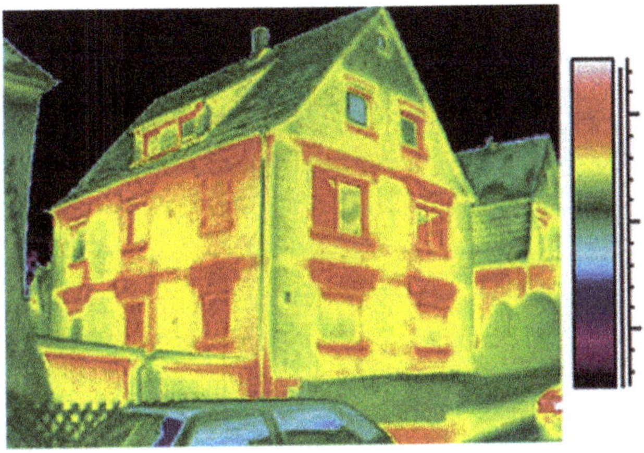

Fig. 4.2: Thermal distribution across a residential building

U is the inverse of R (Thermal Resistance) with SI units of $W/(m^2K)$ and US units of BTU/(h °F ft^2). Another definition of U-value would be, that thermal transmittance is a measure of the speed with which heat is lost through 1 square metre of the element with 1K temperature difference across its faces. The lower the U-value, the better insulation a material can provide against thermal losses. The U-value also incorporates the characteristics of each layer in the element.

4.2.1.1 Thermal Conductivity

The magnitude of heat conducted through a 1 cubic metre of material which has 1K temperature difference between its two opposite faces is known as Thermal conductivity (k). Thermal conductivity is measured in watts per kelvin-metre ($WK^{-1}m^{-1}$, *i.e.* W/(Km). The reciprocal of thermal conductivity is thermal resistivity. Thermal conductivity is imperative when it comes to distinguishing between two materials which could supplement each other in their usage as building fabric. Certain conditions have to be fixed in order for this distinction to be feasible. A general sense of things would tell us that there would be a huge difference in the thermal conductivity of two materials such as brick and polystyrene. However, the difference between a material such as mineral wool and polystyrene would only be highlighted by the figure of thermal conductivity. The table below lists the density and thermal conductivity of common building materials.

Table 4.1: Densities and Thermal Conductance of Common Building Materials

Material	Density (kg/m^3)	Thermal conductivity-k (W/m.K)
Slate	2700	1.9
Sandstone	2100	1.3
Brick Common	1790	0.72
Plaster	1570	0.53
Aerated Block	400	0.08
Carpet	186	0.06
Mineral Wool	25	0.04
Expanded Polystyrene	15	0.033

Courtesy: The Green Building Bible

4.2.2 Thermal Resistance

Practical applications in terms of construction require a term which doesn't get limited by unit measurements which would be more suitable for calculation purposes. For this reason, the term thermal resistance (R) (m^{2}K/W) is used instead of thermal conductance. It is the magnitude of the ability of the thickness of a material to resist the conduction of heat through it. The characteristic is dependent on the thermal conductivity and the thickness of the material. Thermal resistance can be calculated from the following formula:

$$R = t/k$$

[Eqn No. 4.1]

Where:

t is the thickness of the material (m)
k is the thermal conductivity (W/m.K)

In order to reduce the rate of thermal losses from a building, the value of thermal resistance should be as large as possible. This can be achieved by two means:

1. Increasing the thickness of the material
2. Reducing the value of thermal conductivity (k)

4.2.2.1 Air Cavity and Surface Resistance

The presence of air gaps/ air cavities is inevitable in majority of structural elements. In some elements it may be introduced deliberately to provide a barrier for the heat flow path. In other cases it may prevent the passage of rain from the outer surface to the inner surface. The resistance offered by a surface (inner or outer) in the form of still layer that resists the flow of heat is called surface resistance. Heat flow through layers of air occurs by two means – convection and radiation. Emissivity of the surface and direction of heat flow (vertical or horizontal) will determine the amount of thermal resistance that is needed. The thermal resistance of a cavity decreases as the emissivity of the surface increases and vice versa.

4.2.2.2 Calculation of U-value

Construction of an internal envelope with transitions into an external envelope by addition of different layers of structural elements to it, results in the creation of different paths of heat exchanges also known as a thermal bridge. It usually involves a path which comprises of materials which have higher thermal conductivity than the ones whose value of thermal conductivity is lower. Together, the materials create a heat flow path which is considered for the calculation purposes of U-value. According to the Building Regulations code (2002) of UK, the following steps are involved in the calculation of U-value:

1. Identify all heat flow paths through the element
2. Calculate the upper resistance limit (R_{up})
3. Calculate the lower resistance limit (R_{low})
4. Calculate the total resistance (R_T) of the wall from:

$$R_T = \frac{R_{up} + R_{low}}{2}$$

5. Calculate the U-value of wall using:

$$U = \frac{1}{R_T}(m^2K/W) \qquad \text{[Eqn. 4.2]}$$

This can be best illustrated by the help of the figure of a brick and block cavity wall given below:

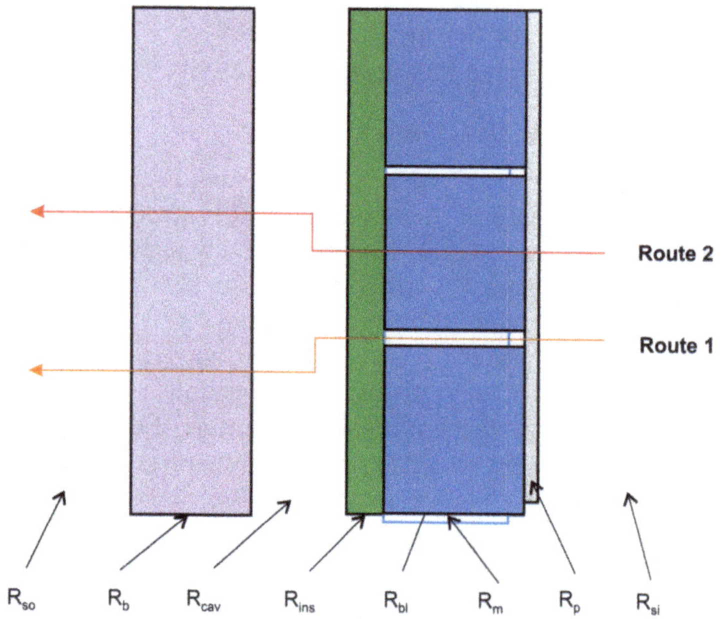

Fig. 4.3: Heat flow paths through a common building section

Step 1: Identify all heat flow paths through an element

Analyzing the figure reveals the two possible paths that heat can take whilst flowing through an element. The thermal conductivity of brick and mortar are fairly similar and so can be considered to be a homogenous layer. The probable paths of thermal resistance are:

(a) Route 1 (TR1) = R_{si}, R_p, R_m, R_{ins}, R_{cav}, R_b, R_{so}

(b) Route 2 (TR2) = R_{si}, R_p, R_{bl}, R_{ins}, R_{cav}, R_b, R_{so}

Where,

R_{si} = Interior Resistance,

R_{so} = Outer Resistance,

R_p = Plaster Resistance,

R_b = Brick Resistance,

R_m = Mortar Resistance,

R_{ins} = Insulation Resistance &

R_{cav} = Cavity Resistance

R_{bl} = Block Resistance

Step 2: Calculate the upper resistance limit

The upper resistance limit maybe calculated using:

$R_{upper} = 1/(FA1/TR1 + FA2/TR2)$

Where,

FA1 = Fractional area of route 1

FA2 = Fractional area of route 2

TR1 = Total thermal resistance of route 1

TR2 = Total thermal resistance of route 2

Determine the total thermal resistances of the two routes by summing individual thermal resistances along the route as shown in the table;

$$TR1 = 0.123 + 0.033 + 0.875 + 2.5 + 0.18 + 0.144 + 0.060$$

$$TR1 = 3.92\ m^2K/W$$

$$TR2 = 0.123 + 0.033 + 0.553 + 2.5 + 0.180 + 0.144 + 0.060$$

$$TR2 = 3.59\ m^2K/W$$

Table 4.2: Values of Thermal Resistance of Each Layer

Layer	Thickness(m)	Thermal conductivity (W/mK)	Thermal Resistance (m^2K/W)
Surface R_{si}	-	-	0.123
Plaster R_p	0.015	0.46	0.033
Mortar R_m	0.105	0.12	0.875
Block R_{bl}	0.105	0.19	0.553
Insulation R_{ins}	0.075	0.03	2.500
Cavity R_{cav}	-	-	0.180
Brick R_b	0.105	0.73	0.144
Surface R_{so}	-	-	0.060

Courtesy: The Green Building Bible

Calculate the upper resistance values by substituting the known values in equation 1:

$$R_{up} = 1/(0.1/3.92 + 0.9/3.59)$$

$$R_{up} = 3.62\ m^2K/W$$

Step 3: Calculate the lower resistance limit

By adding the thermal resistance of each layer, the lower resistance limit. For the block/mortar route:

$$R_{bl/m} = 1/(FA_{bl}/TR_{bl} + FA_m/TR_m)$$

$$R_{bl/m} = 1/(0.9/0.553 + 0.1/0.875)$$
$$R_{bl/m} = 0.574\ m^2W/K$$

Now calculate lower resistance limit using the resistance of the block/mortar layer:

$$R_{low} = R_{si} + R_p + R_{bl/m} + R_{ins} + R_{cav} + R_b + R_{so}$$
$$R_{low} = 0.123 + 0.033 + 0.574 + 2.5 + 0.180 + 0.144 + 0.060$$
$$R_{low} = 3.61\ m^2W/K$$

4 : Calculate the U-value of the element

The U-value is found using;

$$U = 2(R_{up} + R_{low})$$
$$U = 2(3.62 + 3.61)$$
$$U = 0.28\ W/m^2K$$

4.2.3 Heat Loss Equation

Steady state heat loss rate

Heat is lost from the building through the building envelope not only by conduction, convection and radiation but also by the escape of heated air from the building. These are known as fabric and ventilation heat losses. Two mechanisms by which to calculate the speed at which the heat is lost are fabric heat loss rate (Q_f) and the ventilation heat loss rate (Q_v).

$$Q_f = \Sigma(U_x.A_x)\Delta T$$
$$Q_v = 1/3.n.V.\Delta T$$

[Eqn. No. 4.3]

4.2.3.1 Fabric Heat Loss Equation

The first equation shows that the fabric heat loss rate (Q_f) is directly proportional to the area of the element, therefore, any changes to the form that increase the surface area will result in an increased heat loss rate. Fabric heat losses depend on the thermal transmittance of the fabric and the area of envelope through which heat loss can take place. This varies for each individual element and so each element must be treated separately then summed up. The temperature difference between the inside and the outside is the only factor which remains the same of each element. The variables in the heat loss equation represent:

A_x is the area of the element x (window, wall, floor or roof) measured in square metres (m^2).

U_x is the thermal transmittance of element x, also known as the U-value, measured in W/m^2K. Basically this is the heat loss rate in watts through $1m^2$ of the element with 1K temperature difference across its faces.

ΔT is the difference between inside and outside temperature (K).

4.2.3.2 Ventilation Heat Loss Equation

The amount of ventilation heat losses depends upon the temperature of the internal air compared to the outside air (the warmer the air the greater its energy content), and the quantity of it exiting the building. The following variables are used in the equation:

n is the ventilation rate measured as the number of air changes per hour (ac/h). This actually represents the number of times fresh air would completely replace the air in the room in the span of one hour. So, for example, if all of the warm air in a room were to be replaced by cool outside air in an hour, then the ventilation rate would be 1.0 ac/h.

V is the volume of the measured space in cubic meters (m^3). The amount of energy lost will depend on the volume of warm air leaving the building.

ΔT is the difference in temperature between the air inside the building and outside the building, measured in degrees Kelvin (K). A change in temperature of 1K is the same as a change of 1 degree centigrade (°C).

4.2.3.3 Example Calculations

Consider the example of a simple detached house as shown in figure below.

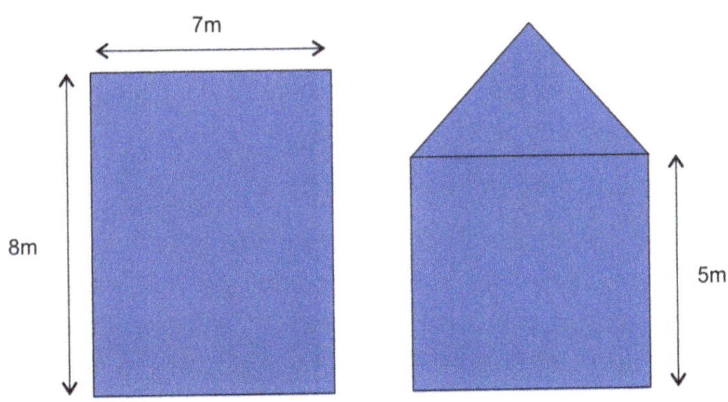

Fig. 4.4: Plan and section of a simple detached house

Table 4.3: Surface Area and U-value of Individual Elements

Element	Area	U-value (W/m^2K)
Floor	56	0.25
Roof	56	0.16
Walls	115	0.35
Window/door	35	2.00

4.2.3.4 Calculating the Fabric Heat Loss Rate

The U-value depicts the heat loss rate through one square metre of a building element for a temperature difference of 1K across its internal and external faces. To determine the heat loss rate through the entire element, the U-value should be multiplied by the area of the element and the actual temperature difference across its faces. For an internal temperature of 25°C and the external temperature of 5°C, temperature difference (ΔT) would be, 25 – 5 = 20°C. For example the fabric heat loss rate through the windows of the example house shown above is calculated from:

$$Q_{windows} = U_{windows} \times A_{windows} \times \Delta T \qquad \text{[Eqn No. 4.4]}$$

Where:

Q = heat loss rate (W)

U = thermal transmittance (W/m^2K)

A = area of element (m^2)

ΔT = air temperature difference between inside and outside (K)

Substituting the given values for the windows from the table into the above equation we get;

$$Q_{windows} = 2.0 \times 35.0 \times 20.0 \text{ (W)}$$

$$Q_{windows} = 1400\text{W}$$

The heat loss rates for all the elements are shown in a table below. The final column shows the percentage contribution to the total fabric heat loss rate made by each element. Even though the windows have the smallest area, they are responsible for over half of the fabric heat loss rate from the building. The table also shows the total value for fabric heat loss rate of 2664.2W.

Table 4.4: Thermal Transmittance, Area, Temperature Difference and Heat Loss Rate for Individual Elements

Element	U	A	ΔT	Q	%
Walls	0.35	115	20	805	30
Floor	0.25	56	20	280	10.5
Roof	0.16	56	20	179.2	7
Window	2.00	35	20	1400	52.5
			TOTAL	**2664.2**	**100**

4.2.3.5 Calculating the Ventilation Heat Loss Rate

The volume (V) of the building is 280 m^3. We will assume some attention has been paid to the draught proofing and the average whole house ventilation rate (n) is 0.85 ac/h. These values can be used in the equation no. 4.3 for ventilation heat loss rate.

$$Q_v = 1/3 \times n \times V \times \Delta T$$

Substituting the known values gives,

$$Q_v = 1/3 \times 0.85 \times 280 \times 20$$

$$Q_v = 1586 \text{ W}$$

The total heat loss rate from the house is $Q_f + Q_v$

i.e., $$Q_{total} = 2664.2 + 1586$$

$$Q_{total} = 4250 \text{ W}$$

4.3 Super-insulation

Buildings that require no external or internal heating systems whatsoever to maintain the thermal comfort levels are enveloped in super-insulation. The BRECSU (Building Research Energy Conservation Support Unit) states three categories of super insulated housing:

1. Zero Heating
2. Carbon Neutral
3. Autonomous

Zero Heating: Commonly called a 'passive house', this type of building would be so highly insulated and draught proofed that all the heating requirements would be met from casual solar gains, lighting, appliances, cooking and occupants bodies. A minimal heating system for severe winters coupled with appliances would be only needed for the sensitive occupants only.

Carbon Neutral: If the electricity supply was completely derived from renewable sources of energy to be supplied to a 'passive house' described above, the house would be considered as carbon neutral. For reducing the net emissions of a house like this, the levels of CO_2 would have to be cut down to zero. This can be achieved by purchasing green credits which are derived from renewable sources themselves. This can also be achieved by consuming fuels during winter months and exporting the same amount of energy from renewable sources produced on site at other times in the year.

Autonomous: Such a system would not only qualify the above two standards but also be disconnected from the mains such as electricity, gas, water and sewage. A grid connection to the electricity is permitted but the house should generate its own electricity, rain and grey water must be collected for filtration and sterilization and sewage must be treated on site.

4.4 Individual Structural Elements

The structural members that would ultimately be responsible for the thermal losses and comfort of the occupants in terms of building fabric and which would be discussed in detail are listed below:

1. Walls
2. Windows or glazing
3. Doors
4. Floors

4.4.1 Walls

The type of wall construction may vary according to the factors given below:

1. The available budget
2. The purpose of the building
3. Availability of materials
4. The loads on the wall and the scale of the building

The basic requirements of a wall apply in each scenario. The wall will fulfill its structural requirements, have a low U-value overall and act as a barrier between wind and rain.

We can divide walls in two categories:

1. Heavyweight
2. Lightweight

Heavyweight walls have large thermal mass and are usually made of masonry, concrete, *etc*. Lightweight walls are made of wooden frames, straw, grass, *etc*. These categories however do not encapsulate modern walling solutions in entirety since the advent of modern materials has transformed the method of construction altogether. Some examples of such materials would be air-filled concrete blocks, hollow polystyrene concrete blocks, *etc*.

4.4.1.1 Heavyweight Walls

Masonry cavity walls are extremely resilient towards external weather factors and do not deteriorate with long periods of exposure to moisture as well. The cavity provides an insulation component to the wall. This maybe enhanced by filling the cavity with an insulating material. The width of the cavity is indirectly proportional to the U-value, therefore, to further lower the U-value, the cavity width should be increased and insulants which have low thermal conductance should be used.

The inner surface of the wall acts as a thermal mass which propagates passive cooling in summer months and maintains it. As the temperature drops overnight, the thermal mass releases the absorbed heat to the exterior and interior parts. Mechanized systems are used to bring the thermal mass at the start of its absorbing cycle.

4.4.1.2 Lightweight Walls

Lightweight walls constitute of a timber frame with either a brick/ masonry or timber as outer skin and plasterboard forming the inner skin. The timber that would be used on the outer skin would be treated or rendered accordingly so as to provide a robust barrier. Timber framed walls with insulants can achieve a U-value of 0.15 W/m^2K or lower.

Some solutions that can be used in conjunction with a timber framed wall are limecrete, rammed earth, straw bale, frass, firbreglass and hemp (hemp lime and hemcrete).

The size of a building determines the structural materials that would be used. For eg., residential buildings may utilize masonry and timber frame walls whereas multi-storeyed commercial buildings would be built using steel and concrete frames. The facades for smaller buildings would be made of brick slips or other cladding material, whereas taller structures would utilize more glass as their choice of external fabric. Industrial structures would use insulated cladding panels or singular metal sheets.

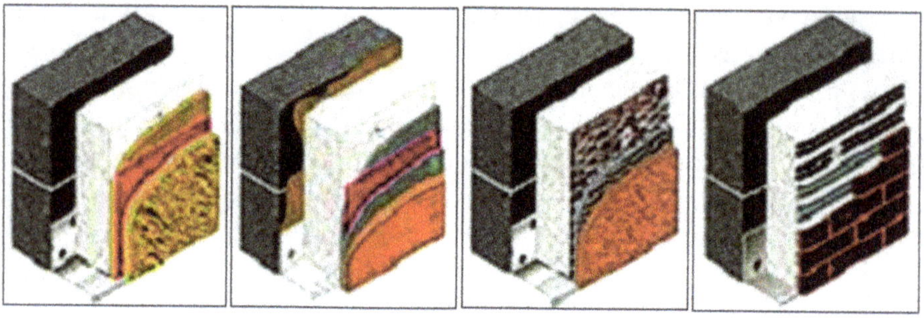

Fig. 4.5: Cross section of different render systems

4.4.1.3 Types & Finishes

Insulation for the external fabric has undergone massive research in development of a lightweight and easily manufactured system that strives to deliver excellent thermal performance. These systems maybe standalone or in layers. Some widely used options are given below:

1. Fibreglass
2. Extruded Polystyrene Foam (XPS)
3. Expanded Polystyrene Boards
4. Cellulose
5. Mineral Wool
6. Polyisocynurate foam slab
7. Cork

Fig. 4.6: Dry dash render system

(Source: www.kilsaran.ie)

4.4.1.4 Render Finishe

Dry Dash: In order to create a relatively cheap and aesthetically pleasing finish, dry render (crushed stone/chips) is applied to a wet surface. The skills required are minimal, therefore, this particular finish is widely adopted. The colours and textures may also be varied in accordance with the requirement.

Scratch plaster: A scratching tool scrapes off 2-3mm of plaster off the surface of the render to lend it a visually appealing feel, once the layer beneath the scraped off layer becomes visible.

Fig. 4.7: Depiction of scratch plaster
(Source: www.skim4u.co.uk)

Rough-cast: The aggregate is mixed-in with cementitious slurry itself. The coated aggregate is then applied directly to the surface which gives off a rough and uneven textured surface due to the spherical shape of the aggregates chosen.

Fig. 4.8: Rough cast render
(Source: www.bosence.co.uk)

Silicone-enriched: Silicon based waterproofing render is relatively new and its innovation brings with it a ready to mix render system that benefits exposed or coastal areas immensely. The application is conventional in its mode with different colours up for choosing.

Smooth/painted finishes: To achieve a smooth rendered finish free of imperfections, the finish is directly applied to a good rendered surface. The surface obtained is reltively free of imperfections and can be produced in a variety of colours.

Textured coatings: These coatings are usually acrylic or silicone bases which are applied by a trowel to a thickness of 1.5-3mm. The coatings induce longevity and water-tightness of the render.

Brick cladding: Brick claddings are usually applied in light-weight construction and form the outer fabric of the structure to give-off a brick wall finish. The brick portion is however, a thin facing that is applied to the outer surface of the insulation.

Fig 4.9: Brick slips (Courtesy: http://www.brickslips.co.uk)

Simulated Stone: Certain polymer renders, offer ability to be painted and coloured in natural stone finish which deliver a strong aesthetic effect without adding any additional burden on application procedures.

Fig. 4.10: Coloured render

4.4.2 Windows

Glass is generally used to provide an unobstructed view from one side to another and also to allow the passage of light. Innovation in glazing units has led to the creation of IGU's (Insulated Glass Units) which not only do the above but also temper the effect of external weather conditions to make the thermal conditions in the interior as comfortable as possible. The glass maybe laminated or tempered according to the requirement whilst varying in thickness to achieve a particular acoustic attenuation if necessary or for security purposes.

The performance of glass can be tweaked through the variation of following parameters:

4.4.2.1 Tinted Glass

Solar heat gain may pose to be a disadvantage in hot/humid conditions especially if the installation utilizes glazing units. In such cases, clear glass is unable to vary the magnitude of direct solar heat gain thus a coat of tint (available in bronze, gray and green) is usually applied to the external surface of the glazing unit to reduce this effect. The coating is treated thermally for adjustment against thermal stresses and breakage.

4.4.2.2 Coated Glass

A polyester or metal coating is applied to improve the insulating as well as acoustic properties of a glazing unit. Such coatings give the glazing a reflective appearance. This coating is a viable option against double glazed units. The film may also be applied to surfaces 1, 2 or 3 (see Fig 4.11) or even a combination of films to further increase the effectiveness of an insulated glass unit as below. Two parameters govern the use of coatings on glazing units – Solar Heat Gain Coefficient (SHGC) and Visible Transmittance. Both of them may be varied in accordance with the requirement as well as climatic conditions.

4.4.2.3 Low-emissivity Glass

A glazing unit with a thin coating (usually metal) in the interior of the unit is known as Low-emissivity glass or Low-e glazing. The coating reduces the Solar Heat Gain Coefficient of the unit by a significant margin.

The position of coating determines the usage of glazing unit. For e.g. in tropical and sub-tropical areas the coating maybe applied to surface 2, whereas in cold climes, the coating would be applied on surface 3, in order to reduce heat loss which would be the opposite of the previous case, in which, the objective is to reduce heat gain. In both cases, the insulating properties of the IGU stay the same.

4.4.2.4 Spacers

Spacers or spacer bars separate the panes of glass in an insulated glazing unit. Metal spacers have proven to be redundant in the overall effectiveness of the glazing unit as a whole since they tend to act as a conductor between the two panes of glass. This results in the formation of condensation which may turn to ice in cold climates. To counter this, the spacers are recommended to be of either structural foam or silicone foam.

4.4.2.5 Efficiency of IGU's

The efficiency of an IGU is primarily determined by the gap between two or multiple panes which may contain air, gas or vacuum. When the space provided is lesser, it results in conductive heat losses (one window pane ends up cooling the other through conduction) whereas in cases where the space is wider, it results in convective heat losses (the heated air gap transmits heat from one pane to the other).

U-value is indirectly proportional to the number of window panes. Standard U-values for single, double or triple glazing units are usually 4.8, 2.8 and 2.1 $W/m^2.K$. Redundancy sets in if more than three panes of glazing are used also resulting in an increase in cost and weight of the unit.

An extremely efficient system of glazing that counters heat loss is Vacuum Insulated Glass (VIG) or evacuated glazing. As the name states, the panes in the glazing units have a vacuum between them. Though difficult to manufacture (like replacing air between the panes with gas), VIG's have almost one fifth the thermal conductivity of conventional insulation. Replacing the air between glass panes with gas, results in a breakthrough technology that has gained momentum in recent times. Gas filled panels (GFP's) are considered as high performance thermal insulating solutions for building applications. The concept is to induce a heavy gas which is more viscous than oxygen or nitrogen. Higher viscous quotient would deliver significant reductions in convective heat transfers. Some of the gases that have found their usage in this operation are argon, krypton and xenon. These gases are non-toxic (which is most important), colourless, odorless and chemically inert. Argon is the most frequently used out of the three due its ease of availability.

4.4.3 Other Transparent Materials

4.4.3.1 GRP

Fibres (usually made of glass) are combined with a fibre reinforced polymer (epoxy, polyester, *etc.*) to form Glass Reinforced Plastic (GRP) or glass fibre reinforced plastic (GFRP). Another name by which it is commonly known as is fibreglass.

Parameters that make GRP a favourable material for building applications are light transmission (about 50%) and its U-value (2.0W/ m^2.K). Apart from these fiberglass is considerable light, strong and a robust material. The fact that it can be molded into variety of shapes is the biggest reason for its vast array of applications. In building applications, fibreglass is primarily used in roofing and door solutions.

4.4.3.2 Aerogel

Aerogel is basically a gel (usually silica based) in which the liquid component has been replaced with gas, resulting in an extremely low-density solid compared to any other porous solids. Also known as 'frozen smoke', aerogel has been used in a plethora of industries such as, aerospace, cosmetic, marine and nuclear weapon development. In building applications, aerogel is used as an insulating material for walling, flooring and roofing solutions. The thermal conductivity for a silica based aerogel has been reported to be almost as low as 0.03 W/m.K.

4.4.3.3 SIP

Structural Insulating Panels (SIP's) are composed of a foam (polystyrene, expanded polystyrene) which is sandwiched between two layers of structural board (sheet metal, plywood sheathes). These panels are extremely versatile in their applicability and are used in foundation, flooring, roofing and walling solutions (in the case of conventional framed buildings). In fact, houses built with SIP's are known to have better load carrying capacity as compared to their stick-framed counterparts.

4.5 Floor

A floor that has been raised above the floor slab to accommodate HVAC and electrical services is known as a suspended or raised floor. These floors are widely used in commercial buildings (server rooms, data banks, *etc.*) due to space constraints and their specialized requirements. The entire services can be routed through the space between the suspended floor and the floor slab without the need for chasing of walls and installing large size conduits.

The usefulness of the suspended floors comes into effect in green structures, when these spaces are used to cool a building through floor-mounted vents. Floor boards make up the larger part of the walking space with cool air flowing below venting at strategically placed location. This practice has been in effect in Europe for almost 40 years now.

The floor is raised on joists which are further provided with an insulation barrier to close air gaps and create a tight fit. A vapour barrier between the insulation and the floorboard prevents the condensation formed underneath the floorboard to move further downward and counters the draughts from the suspended floors to leak into the room above.

Solid Floors

Solid floors are laid directly on top of the floor slab, usually installed in multiple layers all being supported by the ground itself. Screeding is carried out on top to even out the surface. Polystyrene insulations (due to their resistance to compression) are inserted on top of the slab or below, according to the target U-value. It should be noted here that, like every surface with edges, solid floors also require special treatment around its peripheries. The insulation is extended up to the level of the wall insulation which itself is extended, all to prevent a creation of a thermal bridge around the edges. The idea is to increase the thermal mass of the floor by the use of sub-grade insulation or various types of underlay such as carpets.

4.6 Case Study for Thermal Comfort in a Typical Office Building

4.6.1 Thermal Comfort in an Office Building

The graph given below is taken from a study "Modelling and Parametric Study of a Typical Office Segment for Thermal Comfort and Energy Consumption in New Delhi"-(Dhaka 2009)

The variation in energy consumption (y-axis) is evident with respect to change in glazing area and usage of different fabrics (x-axis) for insulation purposes. The parametric analysis was carried out in the GUI based software called IES-VE. For reference, the base case and the add-on fabrics have been illustrated below.

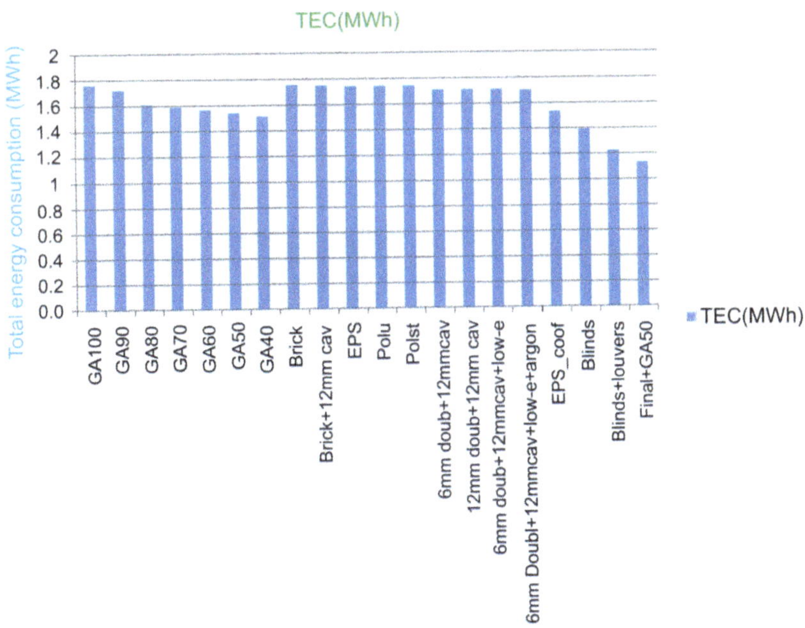

Fig. 4.11: Variation in total energy consumption according to model variation

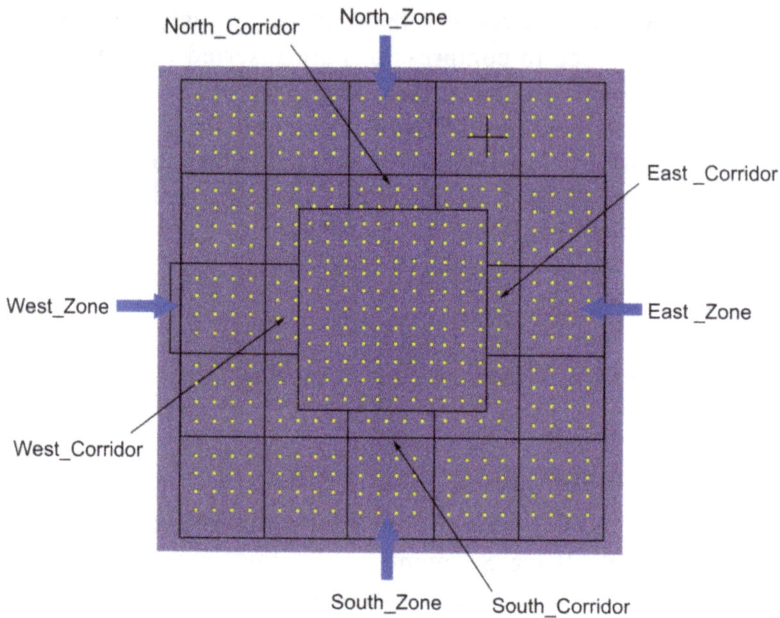

Fig. 4.12: Assumed typical office floor plan

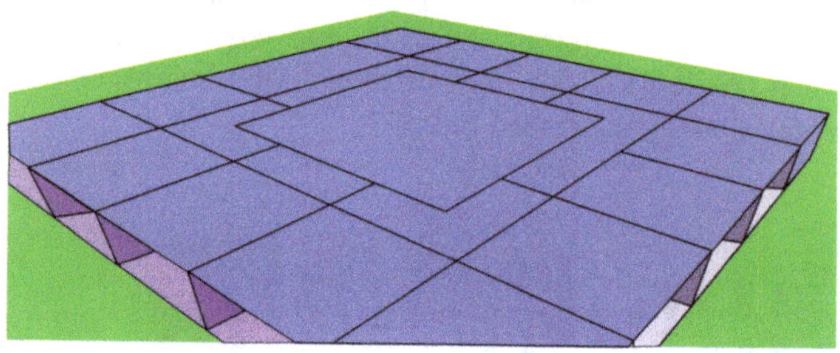

Fig. 4.13: Perspective view of the base model

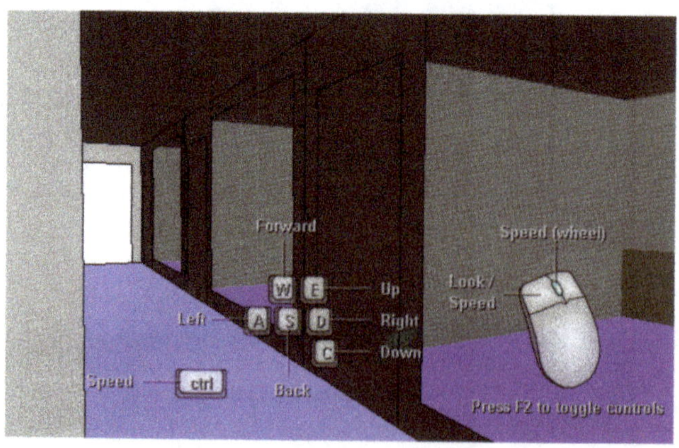

Fig. 4.14: (a) Interior view of base case model

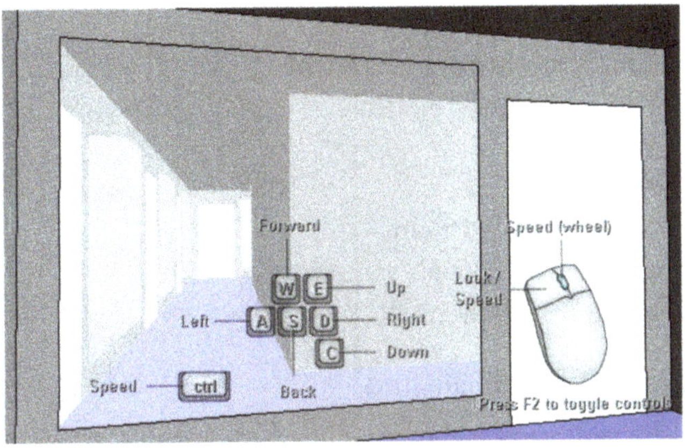

Fig. 4.14: (b) Interior view of base case model

4.7 Base Case Design Considerations

4.7.1 Geometric Properties

The following are the geometric properties of the model:

Dimensions (Entire Floor) = 25 × 25 × 3 m^3

Dimensions (One Cubicle/Office) = 5 × 5 × 3 m^3

Dimensions (One Corridor Segment) = 5 × 2 × 3 m^3

External Façade Area (Entire Floor) = 300 m^2

External Façade Area (Four cubicles) = 60 m^2

Glazing Area (Four Cubicles) = 100%

4.7.2 Design – External Walls

Concrete Thickness = 20 cm

Plaster Thickness (Both sides) = 0.5 cm

Total U-value = 1.97 W $m^{-2}K^{-1}$

4.7.3 Design – Internal Walls

Brickwork Thickness = 10 cm

Plaster Thickness (Both sides) = 0.5 cm

Total U-Value = 2.11 W $m^{-2}K^{-1}$

4.7.4 Design – Roof

Concrete Thickness = 10 cm

Felt/Bitumen Layer = 0.12 cm

Clay Tiles = 5 cm
Plaster Thickness (Inside) = 0.5 cm
Total U-Value = 2.76 W $m^{-2}K^{-1}$

4.7.5 Design – Floor

Concrete Thickness = 10 cm
Synthetic Carpet Thickness = 1 cm
Total U-Value = 2.93 W $m^{-2}K^{-1}$

4.7.6 Design – External Glazing

Vertical Fenestration = 8 mm single clear glass with Metal Frame
Total U-Value = 4.174 W $m^{-2}K^{-1}$

4.7.7 Design – Internal Glazing

Vertical Fenestration = 4 mm single clear glass with Metal Frame
Total U-Value = 4.08 W $m^{-2}K^{-1}$

4.7.8 Design – Others

Door = wooden
Total U-Value = 2.16 W $m^{-2}K^{-1}$
Shading (Internal and/or External) = none

4.8 Building Physics and Operational Design – Assumed Data

1. Occupancy = Each cubicle/individual office has an occupancy of 2
   - (*a*) Maximum Sensible Gain from 1 Person = 60 W/person, therefore for 2 persons = 120 W/person
   - (*b*) Occupancy Profile = 0800 hrs to 1800 hrs (Mon – Fri)
   - (*c*) The produced heat is 100% sensible heat and 0% latent heat
   - (*d*) Corridors have an occupancy of 20 m^2/person
2. Equipment = Each cubicle/individual office has 2 PC's
   - (*a*) Maximum Sensible Gain from 1 PC = 160 W, therefore for 2 PC's = 2 x 160 W = 320 W
   - (*b*) Equipment switched on from = 0800 hrs to 1800 hrs (Mon – Fri)
   - (*c*) The produced heat is 100% sensible heat and 0% latent heat
   - (*d*) There is no equipment profile for corridors

3. Lighting = Fluorescent Lighting
   (*a*) Maximum Sensible Gain from Lighting in Rooms =13 W/m^2 × 0.84 = 10.92W/m^2 and for lighting in corridors = 0.92 W/m^2
   (*b*) Lighting in room and corridors switched on from = 0800 hrs to 1800 hrs (Mon – Fri)
   (c) The produced heat is 100% sensible heat and 0% latent heat
4. HVAC
   (*a*) System = Generic
   (*b*) Simulation Unit Cooling Capacity = Unlimited
   (*c*) Cooling System Profile = 0800 hrs to 1800 hrs (Mon – Fri)
      (*i*) 0000 hrs to 0800 hrs = Off
      (*ii*) 0800 hrs to 1800 hrs = 17-20 °C
      (*iii*) 1800 hrs to 2400 hrs = Off
   (*d*) Generator Fuel = Electricity
   (*e*) Generator Seasonal Efficiency = 0.89
   (*f*) Natural Ventilation = None
   (*g*) Auxiliary Ventilation = None

4.9 Parameters Chosen

1. Glazing Area Variation:
   (*a*) Glaz_100 (100%)
   (*b*) Glaz_90 (90%)
   (*c*) Glaz_80 (80%)
   (*d*) Glaz_70 (70%)
   (*e*) Glaz_60 (60%)
   (*f*) Glaz_50 (50%)
   (*g*) Glaz_40 (40%)
2. Glazing Type Variation:
   (*a*) Cav_6 (6 mm cavity for double glazed)
   (*b*) Cav_12 (12 mm cavity for double glazed)
   (*c*) Sing_6 (Single 6 mm clear glass)
   (*d*) Doub_6 (Double Glazed 6 mm clear glass with a 12 mm cavity)

(*e*) Doub_12 (Double Glazed 12 mm clear glass with a 12 mm cavity)

(*f*) Low-e_Doub_12 (Low-e coating Double Glazed 12 mm clear glass with a 12 mm cavity)

(*g*) Low-e_Doub_12 (Low-e coating Double Glazed 12 mm clear glass with a 12 mm cavity filled with Argon gas)

3. Thermal Mass Variation:
   (*a*) Brick (Brickwork)
   (*b*) Ins_EPS (Expanded Polystyrene Insulation)
   (*c*) Ins_Polu (Polyurethane Insulation)
   (*d*) Ins_Polst (Insulation Polystyrene)
4. Shading Variation:
   (*a*) Int_Shading (Internal Shading) (Blinds)
      (*i*) Weekly Profile = On continuously
      (*ii*) Incident radiation to lower device = 300
      (*iii*) Incident radiation to raise device = 150
      (*iv*) Shading coefficient = 0.6
   (*b*) Ext_Shading (External Shading) (Louvres)
      (*i*) Weekly Profile = On continuously
      (*ii*) Incident radiation to lower device = 300
      (*iii*) Incident radiation to raise device = 150
      (*iv*) Transmission Factors at 15° increments = 1, 0.6, 0.4, 0.2, 0, 0,

4.10 Assumptions

Design Assumptions

1. The air temperature in the room is uniform.
2. The surface temperature of each wall is uniform.
3. The solar radiation, transmitted through the windows of the room is primarily distributed in such a way that:
   (*a*) Each wall receives the same heat flux density by diffuse radiation.
4. The roof of the room receives direct radiation.
5. The radiant temperature of the external environment is equal to the outside air temperature.
6. By means of an ideal temperature control the air temperature in the room can be maintained at a particular level.

4.11 Parameteric Graphical Configurations

4.11.1 External Wall

1. Base Case + HVAC + Internal Gain

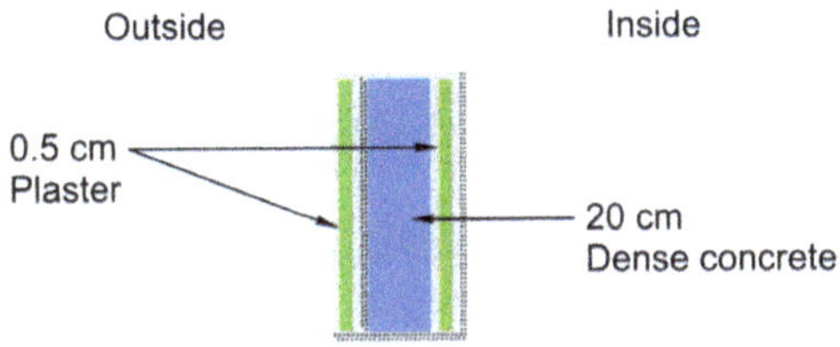

Fig. 4.15: Base case external wall cross section

2. Base Case + HVAC + Internal Gain + Brick Work

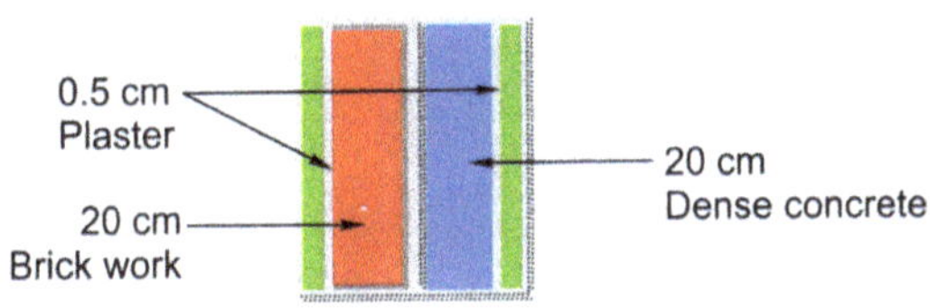

Fig. 4.16. Base case with 20 cm brickwork

3. Base Case + HVAC + Internal Gain + Cavity

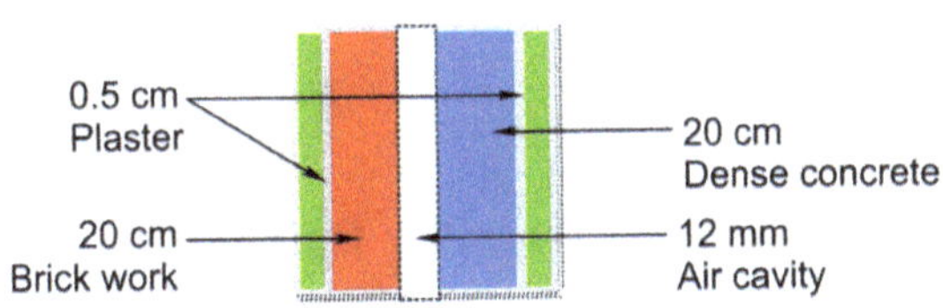

Fig. 4.17: Base case with 20 mm brickwork and 12 mm air cavity

4. Base Case + HVAC + Internal Gain + Insulation (EPS/Polu/Polst)

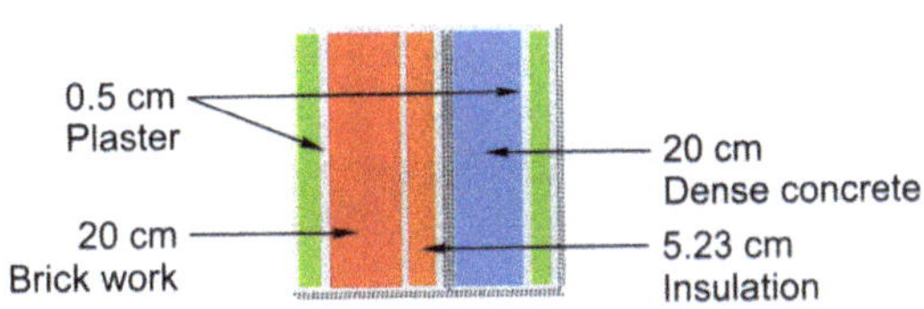

Fig. 4.18: Base case with 20 mm brickwork and 5.23 Cm insulation

4.11.2 Internal Wall

1. Base Case + HVAC + Internal Gain

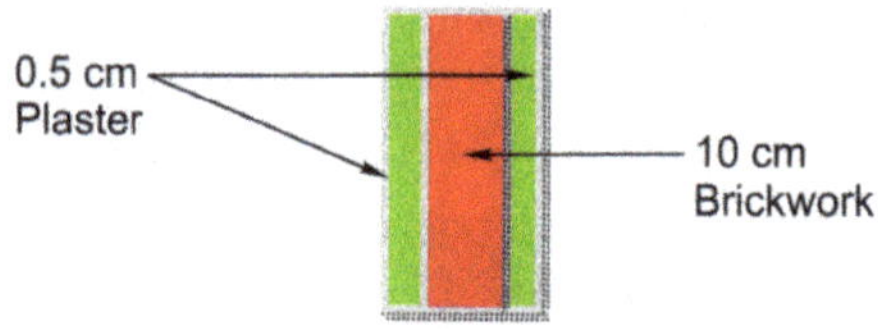

Fig. 4.19: Base case internal wall

4.11.3 Glazing Type

1. Base Case + HVAC + Internal Gain + 6 mm/8 mm Single Glazing clear glass

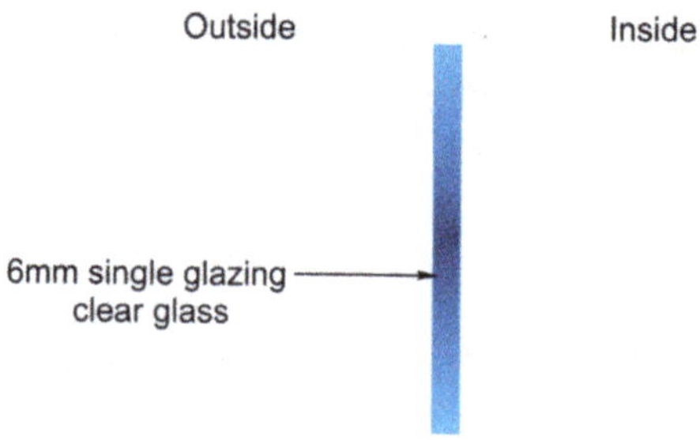

Fig. 4.20: Base case glazing

2. Base Case + HVAC + Internal Gain + 6 mm/12 mm Double Glazing clear glass + 12 mm air cavity/argon gas

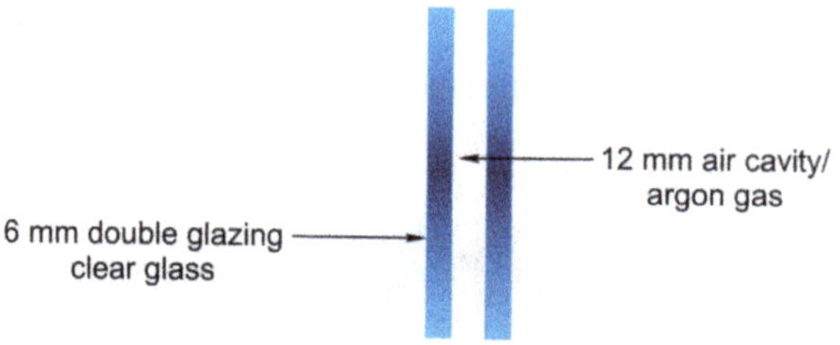

Fig. 4.21: Base case and 6 mm double glazing wih 12 mm air cavity

4.11.4 Roof

1. Base Case + HVAC + Internal Gain

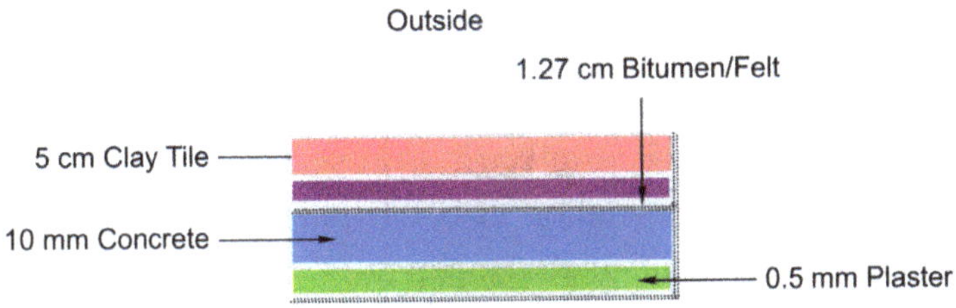

Fig. 4.22: Base case roof slab

2. Base Case + HVAC + Internal Gain + Insulation (EPS)

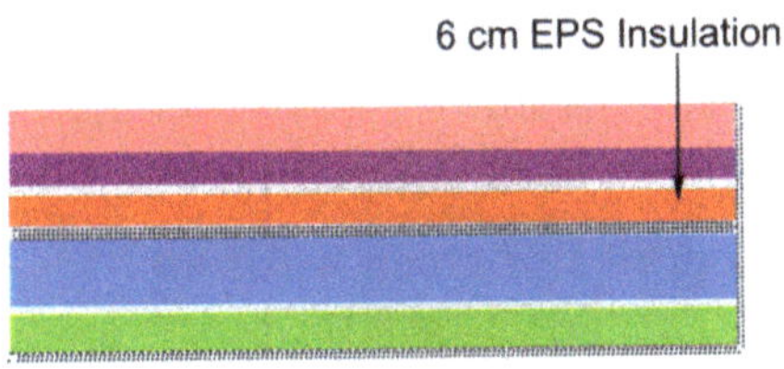

Fig. 4.23: base case and roof slab with EPS insulation

4.12 Green Wall

A wall which maybe standalone or forms a component of a building which is partially or completely covered with any organic or inorganic growing medium is known as a Green wall. The growing medium is usually installed on the exterior face of the wall, however in some cases, for pure aesthetics, the vegetation maybe applied on the interior portion as well. These walls are often referred to as living walls or bio walls as well.

The proof of these walls can be seen in paintings that depict the history of Asia and the Romantic Era (18th Century). Pergolas and arbours, covered in vines embellish these works of art. Climbing plants such as Wisteria were widely used in Asia over bridge railings. Eighteenth century castles saw the cultivation of green facades which typically comprised of walls covered with ivy. The wild vine, with its trademark red autumn leaves became quite the favorite as well. The regular care and maintainence however, took a toll on its growing popularity in the 20th century.

4.12.1 Function of a Green Façade

Urban environments are rife with rising temperatures due to higher population and the added load of systems which sustain a required thermal comfort for that populace. Due to this reason, green walls prove to be extremely effective in such environments to prevent heat build-up. The phenomenon in which heat is absorbed by a thermal mass and is subsequently re-radiated is known as Insolation. It is this phenomenon which needs to be checked to bring down overall temperatures. The vegetative surface of a green wall may increase 4-5 deg C above the ambient temperature, but stays relatively cool. The plants also initiate purificiation of water which maybe polluted after it has been used for various purposes (greywater).

The verticality of green facades proves to be utmost useful for urban environments. They provide a strong aesthetic element to the built vertical face whilst maximizing use of water by allowing less of it to evaporate in comparison to water circulating in horizontal surface.

The two main categories of green walls are:

1. Green Facades
2. Living Walls

Green Facades: These are made up of climbing plants which may either stem from the ground or through a supporting structure on the wall.

Living Walls: The surface of the wall consists of panels which support growth medium. These walls are made up of geotextiles, stainless steel containers, hydroponic systems and the choice of growth medium.

Although green facades are more common in the developed west, Indian studies are also being reported. Patra and Ghosh (2010) have quantified the removal rates of oxides of sulphur by a facade draped with passion flowers. A particular emphasis was placed on the mechanistic details of mass transfer through bulk stomatal cavities. (Quantifying trace gas uptake by passion flower draped facades and roofs by S.Patra and S.Ghosh, Proc.World Green Roof Congress, London, 2010).

4.12.2 Green Roof

A roof enveloped with vegetation or any growing medium maybe termed as green roof or green. The fabric of the roof is provided with various barriers/sheets such as waterproofing, anti-rust, *etc* to make it susceptible to moisture and conditions that may arise due to the growing medium. The speed of urban development has made 'Heat-island' is an inevitable effect which needs isolated micro-measures such as Green roofs to ameliorate it. Roofs covered with vegetation prevent runoff, absorb rainwater, provide insulation and an ideal thermal environment for the building's occupants. Beyond these functional and environmentally conscious benefits, the appearance of a vegetative envelope tends to have a soothing effect on onlookers and occupants alike. It should be noted here, that roofs which have mechanical devices such as photovoltaic panels and/or solar arrays may also be termed as a green roof.

Fig. 4.24: A green façade installed at a Residence
Courtesy:http://greenrooftopgardens.blogspot.in/

4.13 Jaisalmer : A study of Desert Architecture

Jaisalmer is the epitomy of creating a natural urban environment that not only fulfills the needs of its occupants but also creates a buffer against the harsh weather, in the absence of electricity supply, mechanized transport or any modern building technology and severe shortage of water. It is situated in the Thar Desert, Rajasthan (India). The town is home to unique passive design features that counter the harsh desert climate. These features have been practised for generations and are basically a norm of the locals of Jaisalmer.

Fig. 4.25: Model of Jaisalmer town

Socioeconomic status of the occupant is the deciding factor in the type of house in Jaisalmer. There are three types. High income households reside in a structure popularly called as a Haveli. These are built by rich merchants and are usually 4-5 storeys in height. These have rooms and verandahs on all sides of the courtyard and some of them even have basements. The topmost storey has a terrace with parapet walls for safety. Some of the houses might even have more than one courtyard. Middle income households occupy structures which are considered to be the most typical of Jaisalmer with 2-3 storeys to them. The additional room and closed terrace is situated on the upper floor with a balcony facing the street. Low income households occupy houses that are usually single storey with three structures to them. These type of houses are situated on the periphery of Jaisalmer.

4.13.1 Building Fabric Used for Construction

Structures in Jaisalmer have a light-yellowish shade to their facade. This is due to the sandstone which is readily available locally. The stone is such that joints are made accurately without the use of any mortar. Keys are cut into individual stones which fit together or in some cases are held by iron cramps. Low-income households have undressed stone with mud mortar and finishing of mud plaster. Limestone panels are used on the exterior of upper floors as an aesthetic element. These are carved in various patterns to accentuate its natural finish. The latticework is known as Jali in local parlance.

4.13.2 Passive cooling in Jaisalmer

Jaisamer's layout is such that it is the first in the line of significant mesaures that combats harsh weather. The West-North-West to East-South-East orientation of streets encourage the shading of the exterior of any building either by a balcony or sunshades or even buildings on the opposite side. In cases where the the street orientation would be East-West, the summer sun would shine on the south facade from 9:30am to 2:30pm. The solar altitude would be 54°–86° and shading would be further enunciated by small horizontal projections on the south face of the building. The north face would receive solar radiation before 8am and after 4pm in which case the solar altitude would be less than 35°. The building opposite would shade the north face of this building even in case the gap between the buildings is wide. In cases where the street is oriented North-South, during summer season the eastern face would receive solar radiation till 11:30am and the west face after 12:30pm. During this phase the solar altitude would vary from 0° to 79°. In case of a narrow street, the building face would be in shade at all hours except from 10:30am-1:30pm. Therefore, the east-west face would receive no more than an hour of incident solar radiation which is countered by the massive wall.

Fig. 4.26: Narrow alleways

4.13.3 Texture

Texture has been put to ingenious use in Jaisalmer at three levels. The first level, which is at the town scale, the skyline is uneven due to varying parapet heights which leads to mutual shading in the process. The second level involves projections and shading devices which adorn a building's facade. The third level involves the carved portions of the building which act as 'fins'. These act as extensions to the existing structure and induce convective heat transfer to the outside air, in the process cooling down faster than plain surfaces. This concept is widely used in engineering applications where higher rate of heat transfer is a priority. These features, however, are not desirable in cold climates.

Fig. 4.27: Yellow sandstone used with plain cement mortar

A downside of this is that the 'finned' surfaces would absorb heat at a higher rate as compared to a plain surface. Taking this into consideration, one should be careful in analyzing their utility in regard to the building form and orientation. The textured surfaces find their usage in the upper portions on the building which have the propensity to be exposed to sun and that too for longer durations. The lower part of the north and south facades are usually adorned with plain walls along with the entire east and west facades, which are shaded for better part of the day have a plain texture to them as well.

It should be noted that the cooling effect due to the texture of a surface works well only in case of thin walls. Mass-heavy walls made of materials such as brick, stone or mud are self sufficient in reducing solar heat gains and do not require textured surfaces to induct heat loss for them. Wall thicknesses are usually 50 mm for thin walls and 450-600mm for massive walls.

References

1. J E Aronium. 'Climate and Architecture.' Reinhold Publishing Corporation, New York.
2. B M Givoni. 'Climate and Architecture.' Applied Science Publishers, London, UK, 1976.
3. S Jarmul. 'The Architecture Guide to Energy Conservation.' McGraw Hill Book Company, New York, 1980.
4. Koenigsberger, et al. 'Manual of tropical Housing and Building (Part I) : Climatic Design.' Longman Press, India, 1975.
5. C P Kukreja. 'Tropical Architecture.' Tata McGraw Hill Company Ltd, New Delhi, 1978.
6. E Martin. 'Housing Climate and Comfort.' Architectural Press Ltd, UK, 1980.
7. V Olgay. 'Design with Climate.' Princeton University Press, USA, 1973.
8. W F Wagner. 'Energy Efficient Buildings.' McGraw Hill Publishing Company, New York, 1980.
9. D Watson. 'Climatic Design.' McGraw Hill Company, New York, 1983.
10. B Roloff. 'Rajasthan Haveli.'
11. M Fuller. 'Passive Techniques of Heating and Cooling in Buildings.'
12. Fabric and elements - The Green Building Bible.

5

Renewable Energy

Chapter Overview

Renewable Energy (RE) is energy obtained by tapping into the natural energy flow around us. For example, passive solar heating relies on the infra red (IR) part of the electromagnetic (EM) radiation from the sun, and wind power uses the kinetic energy embodied in moving air. Biomass is a form of the chemical energy from the growth of plants following the absorption of sunlight. The main advantages of RE are that it does not rely on a finite fuel reserve and does not emit CO_2 when used. However, whilst it is assumed that biomass is carbon neutral, (since growth of new plants absorbs CO_2 released during combustion of the old ones) there will in fact, be some CO_2 released due to the processing (*e.g.* chipping) of the biomass and also due to transportation of the biomass from where it is grown to where it is needed, the details of which will be discussed in forthcoming sections.

5.1 Solar Energy

5.1.1 Passive Solar Heating

Passive Solar heating is simply the collection of solar energy by a building without using special devices, such as solar panels. However, elements of the building may be modified to optimize solar collection (such as increasing the area of south facing glazing). For a passive solar design to be successful, the following must be considered:

- Availability of solar energy
- Solar collection methods : windows and conservatories
- Thermal storage - since solar energy is unpredictable
- Heat distribution – for instance, solar energy is a south side phenomenon in the northern hemisphere.

The following are the four fundamental areas that cover the above aspects:

Siting: Allow good solar access to the site but make sure buildings that are collecting solar energy themselves do not overshadow nearby buildings.

Form: Extend south facing façade to increase the area for solar collection.

Fabric: Increase the glazed area on the south side and reduce it on the north side.

Cooling: Provide shading devices on windows and thermal mass in the interior to avoid summertime overheating.

5.1.2 Direct and Indirect Gain Systems

Advanced passive techniques are used by architects in building design to achieve thermal comfort conditions in cold climate.

Passive solar heating systems can be broadly classified as:

1. Direct gain systems
2. Indirect gain systems

5.1.2.1 Direct gain

Direct gain happens when sunlight streams through an open window thereby lighting up and warming the interior space. In the cooler mid-latitudes in the northern hemisphere, glazing systems are in place that tap in this direct gain – a glazed window is positioned on the southern side to receive naturally available sunlight over the cold winter months. A double glazing along with drapes reinforces heat retention and reduces heat loss at night times. Such south facing glass facades admit sunlight in and allow it to strike floors and walls which have large thermal storage capacities when they are built with brick, concrete and stone. Such direct gain can utilize about 60-75% of the incoming solar energy received by the windows. The room's thermal mass enables heat storage during the day; at night heat is radiated back into the living space making the room comfortable.

5.1.2.2 Indirect Gain

Indirect gains can be achieved through the use of Trombe walls. These are thermally massive walls with vents in place at the top as well as at the bottom. Trombe walls can be typically built with adobe, masonry as well as concrete. The preferred orientation in the northern hemisphere is the south. The outermost surface is painted black (mimicking a black body) – this ensures maximum absorption. The wall is placed just behind the glazing and has an air gap in between.

Short wave solar radiation is absorbed by the wall during day time and this is stored as sensible heat. The aforementioned air gap thus gets heated. This heat lowers the density of air and enables it to enter living spaces through convection mediated by vents. This is followed by cool air from the rooms replacing the hot air. This process sets up a self-fanning convection current. When the vents are closed at night, the stored heat of the day heats up the living space through conduction and radiation.

Trombe walls are not exclusively a western feature. In fact these are used extensively in Leh and Ladakh – these are very much in place at the Ledeg Hostel in Leh. Buildings that have such walls can maintain an indoor temperature of 15°C even when the outside temperature is markedly colder at –11°C. Wall thicknesses typically range from between 200-450mm with an air gap varying between 50-150mm. The total area of each row of vent is about 1% of the storage wall area. Needless to mention, for a country like India, where the temperature swings are large between summer and winter, the Trombe walls must be shaded suitably for reducing summer gains (Mazria 1979: Levi *et al.* 1983)

5.1.2.3 Water wall

Water walls work on the same principle as Trombe walls and as the name suggests, the thermal storage material is water. This is achieved by constructing a wall made up of water drums that are stacked behind the glazing. As in the case of Trombe walls, they are also painted black. These have the advantage that they have proved to be effective in reducing temperature swings. Heat transfer through water walls is faster than through Trombe walls – this requires a more controlled heat distribution particularly if the gained heat is not immediately required. The obvious choice would be to use water walls in buildings that need to be quickly heated up when the occupants move in to work – schools and offices. The problem of overheating of work places during summers can easily be avoided by use of shading devices.

5.1.3 Sun Spaces

Sun spaces are also known as solar rooms and solariums and have grown naturally following the tradition of designing homes in chilly climes to collect the sun's rays. They bring in warmth and light in the house, cutting down on energy consumption. Often times solar panels are also installed on the roofing. Sun spaces are used in Australia and North America (even India) to grow ornamental and herb gardens much like in a greenhouse.

Such considerations will save a lot of energy for the hotel industry in all major hill stations in India.

Glass and plastic with their little structural strength are clumsy to install. They are brittle and they do not flex before breaking. To have them in place requires additional structural supports increasing the cost considerably.

Within the subcontinent, sloped glazing is hardly recommended except in the Himalayan states. Sloped glazing increases the facades' exposure to weather and may induce material failure.

Instead, vertical glazing is the preferred form for the subcontinent. It can support larger areas of glazing, are easier to install and without the clumsy slant is less leaky. In Rajasthan and other areas, shading a sunspace with roof overhangs have been successfully achieved–vertical glazing lends the desirable warmth in winter and the coolth in summer.

Sidewall glazing is generally not recommended for sun spaces. In the Northern hemisphere if one requires glazing, the preferred direction is the east and not the west. The mellower east facing light filters through the glazing and is a welcome feature as compared to unwanted solar gain during harsh Indian summers between 12 pm to 5 pm. Often a solid insulated wall adjacent to a sun space gives it additional protection from unwanted solar gains in addition to heat losses at night.

5.1.4 Solar Water Heating

Solar water heating systems are generally composed of solar thermal collectors, a fluid system to move the heat from the collector to its point of usage and a reservoir or tank for heat storage and subsequent use (Fig. 5.1). The systems may be used to heat domestic water, swimming pool water, or for space heating. The heat can also be used for industrial applications or as an energy input for other uses such as cooling equipment.

Fig. 5.1: Solar water heater used in Auroville

Residential solar thermal installations can be subdivided into two kinds of systems–compact and pumped systems. Both typically include an auxiliary energy source (electric heating element or connection to a gas or fuel oil central heating system) that is activated when the water in the tank falls below a minimum temperature setting such as 50°C. Hence, hot water is always available. The combination of solar hot water heating and using the heat from a wood stove chimney to heat water can enable a hot water system to work year round in northern climates without the supplemental heat requirement of a solar hot water system being met with fossil fuels or electricity.

5.1.4.1 Technique

In order to heat water using solar energy, a collector is fastened to the roof of a building, or on a wall facing the sun. In some cases, the collector may be free-standing. The working fluid is either pumped (active system) or driven by natural convection (passive system) through it.

The collector could be made of a simple glass toppled insulated box with a flat solar absorber made of sheet metal attached to copper pipes and thoroughly given a black coat of paint. Sometimes the absorber includes a set of metal tubes surrounded by an evacuated (near vacuum) glass cylinder. In some cases, before the solar energy is absorbed, a parabolic mirror is used to concentrate sunlight on the tube.

A simple water heating system would pump cold water out to a collector to be heated, the heated water flows back to a collection tank. This type of collector can provide enough hot water for an entire family. Heat is stored in a hot water tank. The volume of this tank will be larger with solar heating systems in order to allow for bad weather, and because the optimum final temperature for the absorber is lower than a typical immersion or combustion heater. The working fluid for the absorber may be the hot water from the tank, but more commonly (at least in pumped systems) is a separate loop of fluid containing anti-freeze and a corrosion inhibitor which delivers heat to the tank through a heat exchanger (commonly a coil of copper tubing within the tank). The 'drain-back' method has no anti-freezing fluid but is economical to maintain. The piping has sloping pipes to drain water into a tank which is not pressurised but is rather open to the atmosphere. When the pump shuts off the flow reverses and the pipes have enough time to empty prior to freezing. However, for a tropical country like India a conventional heater shown in Fig. 5.1 easily suffices.

In the US, Europe and Australia, often times solar water heaters are used along with a central heating system. Here, solar heat is either concentrated in a tank feeding onto the central heating tank. Alternatively, the solar heat exchanger is placed at a lower level in the tank. Needless to mention, central heating is then used only during night times and during grey and misty winter days. For a sunny country like India, solar water heating for washing and bathing is a very attractive option–the collector water can reach temperatures as high as 60°C. Designs are in place that allow for pressure release mechanisms when excess pressure builds up and likewise a heat dumping mechanism is activated when a threshold temperature is attained.

5.1.4.2 Economics, Energy and System Costs

In warm sunny India, anti-freeze protection is not necessary–a batch-type solar water heater is indeed cost effective. Calculating the payback period should systematically account for the installation and operation costs bearing in mind that input solar energy is entirely free in contrast to India's increasingly expensive electricity.

Solar water heaters can save up to 4.5 tonnes of greenhouse gases annually. Therefore, the Kyoto Protocol has built in a system of offering subsidies to users. These water heaters are robust enough to work for up to 25 years with hardly any maintenance. The payback period is 6 years in most cases. In Indian cities, as well as middle-tier towns, uninterrupted electricity is sometimes a luxury–under such situations solar water heaters can easily offer much needed relief.

Basically a solar water heating system has 3 components–the collection unit, the insulated storage tank and perfectly insulated connecting pipes. The collection unit is a flat-plate collector to convert most of the incident solar radiation into thermal energy: The governing equation known as the Hottel-Wilier collection equation is described by

$$Q_u(\mathrm{Wm}^2) = \alpha_0 T_0\, I(t) - U_L(T_P - T_a) \tag{5.1}$$

Where Q_u is the useful energy,

α_0, T_0 is the absorptivity, transmittivity of the glass cover respectively

U_L is the heat transfer coefficient (=7.5 W/m^2 C at 70°C)

$I(t)$ = total radiation and

$(T_p - T_a)$ is the difference in temperature between the plate and the surroundings.

Fabricators of solar water heaters can easily work out the temperature of the heated water from the above equation with known values of the supplied input parameters such as the absorptivity and transmittivity.

5.1.5 Solar Concentrators

Auroville's solar dish power generating system is an Indian triumph showcasing the veracity of a high temperature solar concentrator. This standard design combines an excellent concentrator and an efficient receiver with a high performance heat engine. A solar dish / Stirling electric power generation system generally has a solar dish to power electric power efficiency of the order of 30%.

A dish/Stirling system consists of the following components:

- A sun-tracking system rotates the solar concentrator about two axes to keep its optical axis pointing directly toward the sun. The concentrator's shape allows reflecting the sun's rays into a cavity receiver located at the focal spot.
- The cavity receiver absorbs the concentrated solar energy. Thermal energy then heats the working gas in the Stirling engine.
- The Stirling engine converts the heat into electricity.

The size of the solar collector (*i.e.* concentrator) of dish/Stirling systems is determined by the power output desired at maximum insolation levels (nominal 1000 Wm^{-2}) as well as the collector and power-conversion efficiencies. With current technologies a 5 kW_e dish/Stirling system requires a dish of approximately 5.5 m in diameter whereas for a 25 kW_e system the installation of a dish of about 10 m in diameter is necessary.

The ideal shape for the reflecting surface of a solar concentrator is a paraboloid (Fig. 5.2). In practice, however, it is often easier to fabricate multiple spherically shaped surfaces. Spherically shaped surfaces also concentrate solar radiation when the focal spot is many mirror diameters away from the reflecting surface (*i.e.* the mirror is only slightly curved), because the parabola can be well approximated in such a small section. Here the solar intensity falling over a large surface is concentrated to a smaller area or receiver. The concentration is achieved by placing a reflector behind the receiver.

5.1.5.1 Parabolic Trough

This type of collector is generally used in solar power plants. A trough-shaped parabolic reflector is used to concentrate sunlight on an insulated tube (Dewar tube) or heat pipe, placed at the focal point, containing a coolant which transfers heat from the collectors to the boilers in the power station.

PARABOLIC TROUGH COLLECTOR

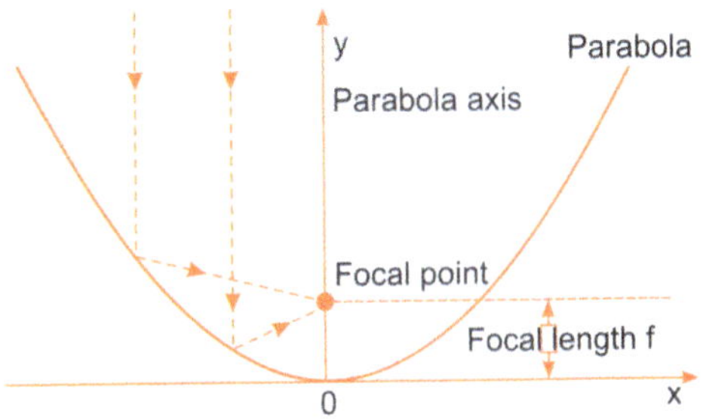

Fig. 5.2: Parabolic solar collector used in solar kitchen in Auroville

5.1.6 Photovoltaic Panels

The origins of the word 'photovoltaic' is Greek–'phos' meaning light and 'voltaic' meaning electricity. The term has a continuous English usage since 1849.

Becquerel first discovered the effect but the first solar cell was built by Charles Fritts who coated Selenium with an exceedingly thin layer of gold at the junctions. The device, however, was only 1% efficient. The modern age of solar power technology arrived in 1954 when Bell Laboratories, experimenting with semi-conductors, accidentally found that silicon doped with certain impurities was very sensitive to light. The efficiency at this time was still around 6 percent. It was only in 1970 that the first highly effective GaAs heterostructure solar cells were created by Zhores Alferov and his team in the USSR.

When photons hit the solar panel they are absorbed by the semi-conductor and electrons are knocked out. This flowing negative charge produces electricity. There is a complimentary positive charge–called holes–with a flow direction opposite to those of electrons. With this simple flow process an array of solar panels can convert the incoming solar energy to a usable amount of DC electricity. Alternatively, the DC current entering into an inverter is converted to either 120 or 240 volt AC depending on the country. The AC power is fed into the utility panel of the house and the electricity is distributed to the appliances therein. The unused electricity is siphoned to other facilities.

When a photon hits a semi-conductor either it passes straight through it (for low energy photons), or it reflects off the surface. The third possibility is that it is absorbed by the semi-conductor which either generates heat or electron-hole pairs–this only happens when the photon energy is higher than the semi-conductor band gap. Note that if a photon has an integer multiple of band gap energy, it can create more than one electron-hole pair. However, this effect is usually not significant in solar cells. The "integer multiple" part is a result of quantum mechanics and the quantization of energy.

When a photon is absorbed, its energy is given to an electron in the crystal lattice. Usually this electron is in the valence band, and is tightly bound in covalent bonds between neighbouring atoms, and hence unable to move far. The energy given to it by the photon "excites" it into the conduction band, where it is free to move around within the semi-conductor. The covalent bond that the electron was previously a part of now has one fewer electron – the hole. The presence of a missing covalent bond allows the bonded electrons of neighbouring atoms to move in the "hole", leaving another hole behind, and in this way a hole can move through the lattice. Thus, it can be said that photons absorbed in the semi-conductor create mobile electron hole pairs.

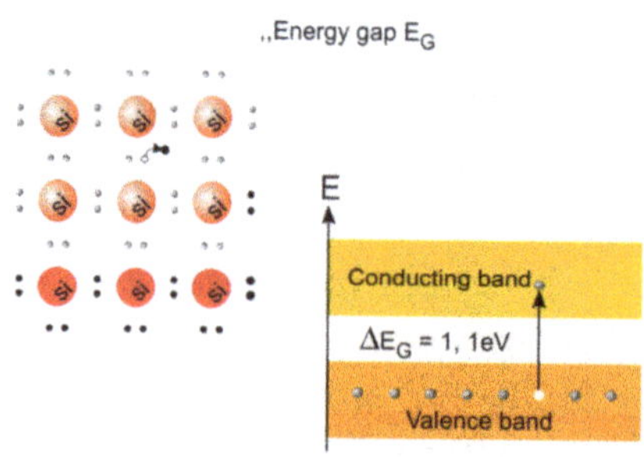

Fig. 5.3: Definition sketch of semi-conductors showing valance and conduction bands

For a transition of the electron from the valence into the conduction band, the knocking photon must have a greater energy than that of the band gap. If one assumes a black body temperature of 6000 K for the sun, the arriving photons generally achieve this, enabling them to be absorbed by the solar cell. However, the energy difference between the arriving photons and this silicon band gap is converted to heat mediated by phonons (a unit of lattice vibration) rather than to usable electricity. (Fig. 5.3).

There are two main modes for charge carrier separation in a solar cell–the drift of carriers, driven by an electrostatic field established across the device and the diffusion of carriers from zones of high carrier density to zones of low carrier density (mediated by a gradient of electrochemical potential).

In the widely used p-n junction solar cells, the dominant mode of charge carrier separation is by drift. However, in non p-n junction solar cells (third generation solar cells using dye and thin film polymers), a general electrostatic field has been confirmed to be absent, and the dominant mode of separation is via charge carrier diffusion.

5.1.6.1 The p-n Junction

The most commonly known solar cell is configured as a large-area p-n junction made from silicon. As a simplification one can imagine bringing a layer of n-type silicon into direct contact with the layer of p-type silicon. In practice, p-n junctions of silicon solar cells are not made in this way, but rather, by diffusing an n-type dopant into one side of a p-type wafer (or vice versa).

If a piece of p-type silicon is placed in intimate contact with a piece of n-type silicon, then a diffusion of electrons occurs from the region of high electron concentration (the n-type side of the junction). When the electrons diffuse across the p-n junction, they recombine with holes on the p-type side. The diffusion of carriers does not happen indefinitely. The electric field results in a charge imbalance on either side of the junction much like a diode promoting unidirectional flow across the junction. Electrons pass from the end side of the junction–this region where the electrons have diffused across the junction does not carry mobile charges and is therefore known as the depletion region or the space charge region.

5.1.6.2 Solar Cell Efficiency Factors

Determination of the Maximum Power Point: A solar cell may operate over a wide range of voltages and currents. By increasing the resistive load on an irradiated cell continuously from zero (a short circuit) to a very high value (an open circuit), one can determine the maximum power point that maximizes the voltage times the current, *i.e.*, the load at which the cell can deliver maximum electrical power at that level of irradiation. The maximum power point of a photovoltaic system is therefore, directly dependent on the level of incident radiation. Larger systems often use a maximum power point tracker which tracks the instantaneous power transfer by continually measuring voltage and current–this information is used to concomitantly adjust the load with a maximum power transference independent of lighting variations.

Energy conversion efficiency: A solar cell's energy conversion efficiency (η, "eta"), is the percentage of power converted (from the absorbed light to electrical energy) and collected, when a solar cell is connected to an electrical circuit. This term is calculated using the ratio of P_m divided by the input light irradiance under "standard" test conditions (E, in W/m^2) and the surface area of the solar cell (A_c in m^2).

$$\eta = P_m/(E.A_c) \tag{5.2}$$

At solar noon, on a clear March or September equinox day, the solar radiation at the equator is about 1000 W/m^2. Hence, the "standard" solar radiation (known as the "air mass 1.5 spectrum") has a power density of 1000 watts per square meter. Thus, a 12% efficiency solar cell having 1 m^2 of surface area in full sunlight at solar noon at the equator during either the March or September equinox will produce approximately 120 watts of peak power.

Quantum efficiency: This refers to the percentage of absorbed photons producing electron-hole pairs. This is only intrinsic to the light absorbing material and not to the cell as a whole–the latter is relevant for thin film solar cells. At the outset it must be stated that QE is not the same as energy conversion efficiency since it does not inform us about the power extracted from the cell.

Solar cell efficiencies vary from 6% for amorphous silicon-based solar cells to 42.8% with multiple-junction research lab cells. Solar cell energy conversion efficiencies for commercially available multicrystalline Si solar cells are around 14-16%. The highest efficiency cells have not always been the most economical – for *e.g.* a 30% efficient multijunction cell based on exotic materials such as gallium arsenide or indium selenide and produced in low volume might well cost one hundred times as much as an 8% efficient amorphous silicon cell in mass production, while only delivering about four times the electrical power (Fig. 5.4).

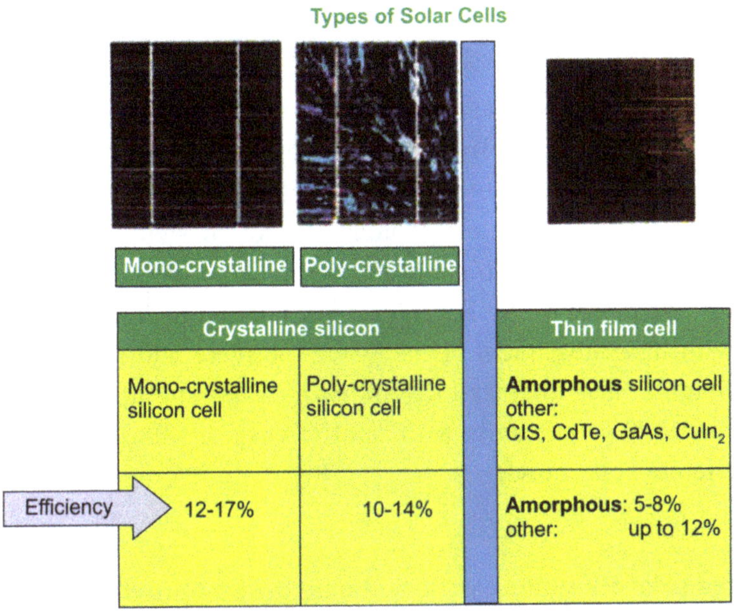

Fig. 5.4: Types of solar cells (Courtesy Univ. of Aachen-SMBS VIT Univ)

To make practical use of solar-generated energy, the electricity is most often fed into the electricity grid using inverters (grid-connected PV systems); in standalone systems, batteries are used to store the energy not immediately required.

5.1.7. Case Studies: Auroville Solar Collector and other Indian Examples

The earliest living huts in Auroville, built as pentagonal capsules on stilts, were made out of thatch material and had no running water. The solar water heaters made for these homes were portable and shaped like a bread box. A bucket was used to feed water into it. The solar heaters provided 60 litres of hot water in the evenings for each square metre of single-glazed collector. They were all hand built and cost U.S.$15.00 each. Although many of these still work, the community has supplemented itself with natural circulation water heaters commercially manufactured in India.

The community takes recourse to solar drying for food processing. Free convection air drying systems were fabricated for melting jaggery, drying spices and organic foods, wood seasoning and the curing of ferro-cement components.

5.1.7.1 Solar Cooking

Box type cookers with a single mirror, designed to Indian Standard specifications are now manufactured for use within the community. A large system capable of delivering community meals is now operational. It has a large ferro-cement bowl with prefabricated reflector segments and a tracking boiler on a computer controlled swinging boom. The bowl with a 15 metres diameter, has an accuracy of 15 minutes of arc. This can be improved to 3 minutes by positioning individual mirror strips with laser alignment. This system delivers 500 kg of steam per day through a heat transfer loop and works as a top up boiler to cook meals for 1,000 people daily with an efficiency of 43%. It is the world's only concrete solar bowl.

5.1.7.2 Solar PV Pumps

Clustered solar PV pumps are most numerous in Auroville as compared to the rest of India. They use the common configuration–the 960 wp module directly linked to a 1 hp centrifugal surface pump. The foundations, installations, module supports, electronics and wiring were all done in-house. Auroville has installed a thousand solar pumps across many states of India and was awarded the Ashden Prize. Fig. 5.5 shows solar powered houses in Auroville.

Fig. 5.5: Solar Powered houses designed with Auroville Technology

5.2 Wind Energy

5.2.1 General Circulation

The sun is the source of all energy received by the earth's atmosphere. The spherical shape of the earth is responsible for the unequal absorption of sun's energy. Without the transport of heat by the atmosphere and the oceans from the equator to the poles, temperatures would be several tens of degrees colder at the poles and warmer at the equator. Upper level winds strongly influence winds near the surface, where most diffusion problems occur (Fig. 5.6).

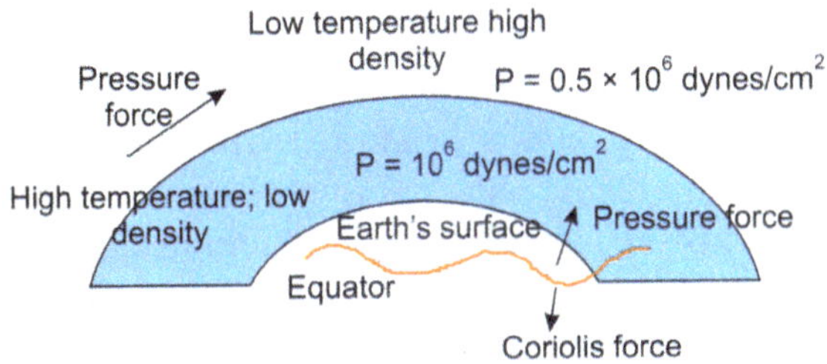

Fig. 5.6: Definition sketch of large scale forces affecting wind flow on earth

The equation of state for the atmosphere is given by:

$$p = \rho RT \tag{5.3}$$

Where R is the gas constant (0.287×10^7 ergs g^{-1})

Equation (5.3) implies that the density must be less at the equator than at the poles.

The hydrostatic equation yields:

$$dp/dz = -\rho g \; (g = 989 \text{ cm/s}_2) \tag{5.4}$$

This suggests that the pressure decreases with height faster at the poles than at the equator. It follows that, if ρ is constant at sea level, then any other constant pressure surface aloft will slope downward from the equator to the pole.

The equation of motion says that air will first be accelerated towards the poles along the upper constant pressure surface:

$$dv/dt = -g(\delta z/\delta y)|p - fu \tag{5.5}$$

where
u = easterly component of the wind speed
v = northerly component of the wind speed
y = northerly coordinate axis
f = Coriolis parameter, which is equal to two times the earth's rotation times the sine of the latitude.

f ~ 10^{-4} s^{-1}, the Coriolis force arises as a result of the earth's rotation, which constantly displaces a Cartesian coordinate system fixed to the surface. In the northern hemisphere this is directed to the right where the poleward force is balanced by the Coriolis towards the equator. For upper level Geostrophic Winds, dv/dt = 0 and this happens when the pressure gradient force exactly balances the Coriolis force. Upper level winds affect wind speeds lower down in the boundary layer and should be considered when designing wind farms.

5.2.2 Vertical Structure and Stability

The planetary boundary layer (PBL), also known as the atmospheric boundary layer (ABL), or peplosphere, is the lowest part of the atmosphere and its behaviour is directly influenced by its contact with a planetary surface. It responds to surface forcings in a timescale of an hour or less. In this layer, physical quantities such as flow velocity, temperature, moisture *etc*, display rapid fluctuations (turbulence) and vertical mixing. Above the PBL, in the "free atmosphere" the wind is approximately Geostrophic (parallel to the isobars) whilst within, it is affected by the surface drag and turns across the isobars. The free atmosphere is usually non turbulent, or only intermittently turbulent.

5.2.2.1 Stability Types

Meteorologists distinguish three states of the atmospheric surface layer: unstable, neutral and stable.

1. If ($\rho_p < \rho_e$ or $T_p > T_e$), then the parcel is accelerated upward (unstable). $\partial T_e/\partial z < -0.98°C / 100$ m
2. If ($\rho_p > \rho_e$ or $T_p < T_e$), then the parcel is accelerated downward (stable) $\partial T_e/\partial z > -0.98°C / 100$ m
3. If ($\rho_p = \rho_e$ or $T_p = T_e$), then the parcel continues at its original speed (neutral) $\partial T_e/\partial z = -0.98°C / 100$ m.

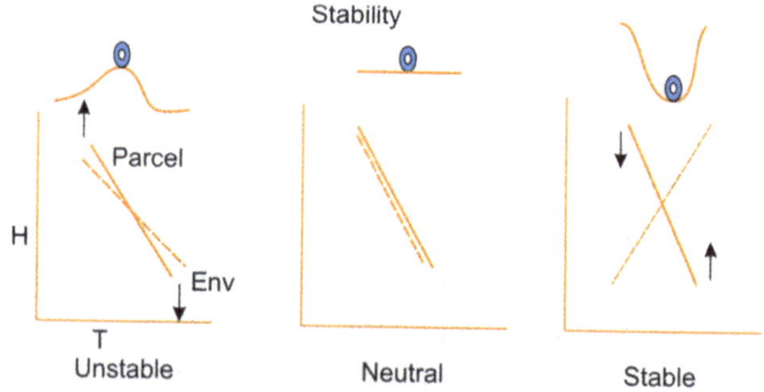

Fig 5.7: Definition sketch for states of stability

One can expect an unstable state at 100m above the surface on a sunny day; a neutral state on a windy or cloudy day and a stable layer near the surface at night.

After sundown, the wind gradient near the surface increases, with the increasing stability, atmospheric stability occurring at night with radiative cooling tends to contain turbulent eddies vertically, increasing the wind gradient. The magnitude of the wind gradient is largely influenced by the height of the convective boundary layer and this effect is even larger over the sea, where there is no diurnal variation of the height of the boundary layer as there is over land. In the convective boundary layer, strong mixing diminishes the vertical wind gradient.

Typically, due to aerodynamic drag, there is a wind gradient in the wind flow just a few hundred meters above the earth's surface – the surface layer of the planetary boundary layer. Wind speed increases with increasing height above the ground, starting from zero due to the no-slip condition. Flow near the surface encounters obstacles that reduce the wind speed, and introduce random vertical and horizontal velocity components at right angles to the main direction of flow. This turbulence causes vertical mixing between the air moving horizontally at one level and the air at those levels immediately above and below it, which is important in dispersion of pollutants and in soil erosion. *(Hanna et al., 1982).*

Wind Velocity Profiles in Dependence of Ground Characteristics

$$\bar{v}_H = \bar{v}_{10} \left[\frac{H}{10\,m} \right]^g \ [m/s]$$

$\bar{v}_H$ = *annual mean velocity in height H* $[m/s]$
$\bar{v}_{10}$ = *annual mean velocity in height* 10 $[m/s]$
H = height $[m]$
g^* = exponent [1]

Description of terrain	Exponent g*
Open country with few and low obstacles, (flat grass and farm land with few trees, prairies, coastline, flat islands, deserts)	0.16
Terrain with uniformally strewn obstacles of 10 to 15 m height (settlements, small cities, woods, shrubberies, small fields with bushes, trees or hedges)	0.28
Terrain with high and non uniformly strewn obstacles (centres of big cities, highly uneven terrain with many high obstacles like trees etc)	0.40

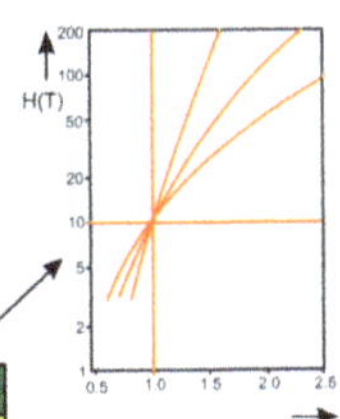

Influence of ground characteristic on the wind velocity profile

Boundary layer exponent for different ground characteristics (Molly, 1978)

Fig. 5.8: Empirical characterization of the surface layer wind profile over differing surfaces (Courtesy: University of Aachen-SMBS VIT University)

The reduction in velocity near the surface is a function of surface roughness, so wind velocity profiles are quite different for different terrain types. Rough, irregular ground, and man-made obstructions on the ground, retard movement of the air near the surface, reducing wind velocity. Because of low surface roughness on the relatively smooth water surface, wind speeds do not increase as much with height above sea level as they do on land (Fig. 5.8).

Over a city or rough terrain, the wind gradient effect could cause a reduction of 40% to 50% of the geostrophic wind speed aloft; while over open water or ice, the reduction may be only 20% to 30%.

For engineering purposes, the wind gradient is modelled as a simple shear exhibiting a vertical velocity profile varying according to a power law with a constant exponential coefficient based on surface type. The height above ground where surface friction has a negligible effect on wind speed is called the "gradient height" and the wind speed above this height is assumed to be a constant called the 'gradient wind speed". For example, typical values for the predicted gradient height are 457 m for large cities, 366 m for suburbs, 274 m for open terrain, and 213 m for open sea.

Although the power law exponent approximation is convenient, it has no theoretical basis. When the temperature profile is adiabatic, the wind speed should vary logarithmically with height, measurement over open terrain in 1961 showed good agreement with the logarithmic fit up to 100 m or so (within the surface layer), with near constant average wind speed up through 1000 m.

The shearing of the wind is usually three-dimensional, that is, there is also a change in direction between the 'free' pressure-driven geostrophic wind and the wind close to the ground. This is related to the Ekman spiral effect. The cross-isobar angle of the diverted ageostrophic flow near the surface ranges from 10° and over open water, to 30° over rough hilly terrain, and can increase to 40°-50° over land at night when the wind speed is very low.

The Log wind profile is a semi-empirical relationship used to describe the vertical distribution of horizontal wind speeds above the ground within the atmospheric surface layer. The relationship is well described in the planetary boundary layer literature. The logarithmic profile of wind speeds is generally limited to the lowest of 100 m of the atmosphere (*i.e.*, the surface layer of the atmospheric boundary layer).

In the free atmosphere, geostrophic wind relationships should be used. The equation to estimate the wind speed (u) at height z (meters) above the ground is:

$$u(z) = (u^*/k)\ln(z/z_0) \tag{5.6}$$

Where u^* is the friction (or shear) velocity (ms^{-1}), k is von Karman's constant (~0.41), z_0 is the surface roughness (in meters).

Roughness length (z_0) is a corrective measure to account for the effect of the roughness of a surface on wind flow, and is between 1/10 and 1/30 of the average height of the roughness elements on the ground. Over smooth open water, expect a value around 0.0002 m, while over flat, open grassland z_0 ~ 0.03 m, cropland ~ 0.1-0.25 m and brush or forest ~ 0.5-1.0 m (values above 1m are rare and indicate excessively rough terrain).

The log wind profile (Fig. 5.9) is generally considered to be a more reliable estimator than the wind profile power law, which is commonly used when neutral conditions are assumed and roughness information is not available. The logarithmic law is strictly valid only in neutral conditions, *i.e.*, when the effect of buoyancy on turbulence is low compared to the effect of wind shear. In such conditions the temperature profile in the surface layer will be close to adiabatic (*i.e.*, $dT/dz = -9.8$ C km^{-1})

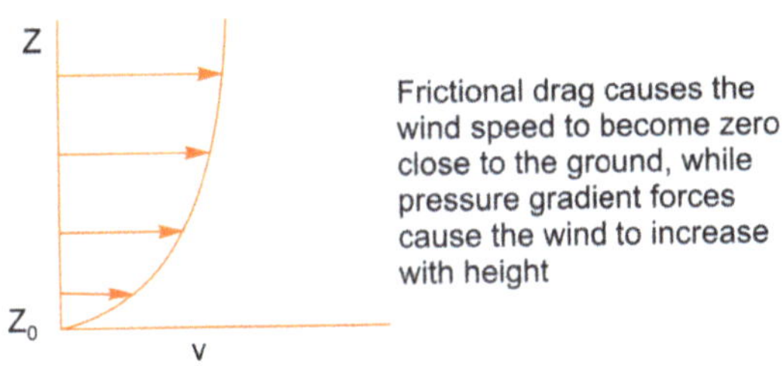

Fig. 5.9: A logarithmic wind profile

5.2.3 Wind Direction, Wind Forces and Wind Velocities

Wind direction is usually reported in cardinal directions or in azimuth degrees. There are a variety of instruments used to measure wind direction, such as the windsock and wind vane. Both of these instruments work by moving to minimize air resistance. The way a weather vane is pointed by prevailing winds indicates the direction from which the wind is blowing. The larger opening of a windsock faces the direction that the wind is blowing from; its tail with the smaller opening points in the direction the wind is blowing.

Modern instruments used to measure wind speed and direction are called anemometers and wind vanes respectively. These types of instruments are used by the wind energy industry, both for wind resource assessment and turbine control.

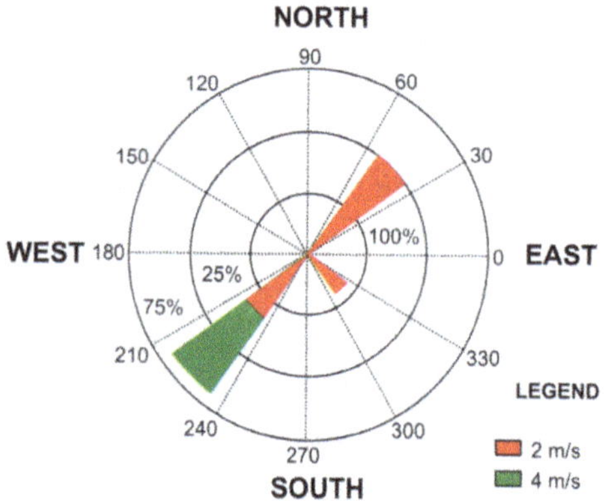

Fig. 5.10: A typical wind rose over Cuddalore in Tamil Nadu, India

A wind rose is a graphic tool used to give a succinct view of how wind speed and direction are typically distributed at a particular location (Fig. 5.10). Using a polar coordinate system of gridding, the frequency of winds over a long time period are plotted by wind direction, with colour bands showing wind ranges. The directions of the rose with the longest spoke show the wind direction with the greatest frequency.

Wind speeds are highly variable and their distribution can be reasonably expressed in terms of the Weibull distribution:

$$P(V) = (k/V)(V/C)^{k-1} \exp\{-(V/C)^{k}\} \tag{5.7}$$

P(V) is the frequency of occurrence of wind speed V,

C is the scale parameter or characteristic wind speed and

k is the shape parameter.

The cumulative Weibull distribution P(V) gives the probability of the wind speed exceeding the value V.

The Beaufort Wind Scale (Fig. 5.11) informs one that the average wind speed also contains a set of wind speed ranges and gives a description for each category. Additionally, it provides an indication of what the sea looks like at that scale with estimated wave height, along with an observation of what the wind will do on land.

Table 5.1: Wind Classification Schemes

Wind Forces Beaufort	Wind Velocity		Effects
	(m/s)	(knots)	
0 calm	0 ... 0.2	0 ... 1	smoke moves vertically
1 very low	0.3 ... 1.5	1 ... 3	smoke moves obliquely
2 low	1.6 ... 3.3	4 ... 7	air flow just sensible
3 weak	3.4 ... 5.4	8 ... 11	leaves movement on trees
4 moderate	5.5 ... 7.9	12 ... 15	spring movement on trees
5 fresh	8.0 ... 10.7	16 ... 21	branch movement on trees
6 strong	10.8 ... 13.8	22 ... 27	howling of wind
7 stiff	13.9 ... 17.1	28 ... 33	trees movement, precipitating waves
8 stormy	17.2 ... 20.7	34 ... 40	tree trunks bow, difficult walking
9 tempest	20.8 ... 24.4	41 ... 47	roof tiles fall down
10 heavy tempest	24.5 ... 28.4	48 ... 55	trees fall over
11 almost hurricane	28.5 ... 32.6	56 ... 63	destroying effects of heavy kind
12 hurricane	32.7 ... 36 .9	64 ... 71	walls fall over, general destruction

Wind Direction: direction from where the wind blows

5.2.4 Wind Energy Conversion Systems

Fig. 5.11 below shows a wind turbine in Auroville, Tamil Nadu. There are more common designs which are discussed in this section.

Fig. 5.11: Wind turbine at auroville placed above the canopy line

The main constituents of a typical modern wind turbine are a rotor with three blades, hub, nacelle and tower. The rotor is upwind of the tower, *i.e.* faces into the wind and rotates in a vertical plane. However, the more commonly used type is the horizontal axis wind turbine. The power present in the wind is

$$P_w = \tfrac{1}{2}\,\rho\pi R^2 u^3 \qquad (5.8)$$

Where ρ is the density, R is the rotor radius and u is the wind speed. It is proportional to the area of the rotor and the wind speed cubed (Fig. 5.12). The power extracted from the wind by the wind turbine rotor is

$$P_R = \frac{1}{2} \rho \pi R^2 C_p u^3 \tag{5.9}$$

C_p is the power coefficient, defined by the ratio of power extracted by rotor to the power and is dependent on the tip speed ratio and the angular rotational speed of the rotor.

In addition, on turbines that can pitch the blades about their longitudinal axes, C_p depends on the blade pitch angle as well. Not all the power in the wind can be extracted by the rotor. The wind speed over the rotor would need to be zero, which would be contradictory as it would imply that the extracted power is also zero. Consequently, the value of C_p must be less than one. The theoretical maximum is the Betz limit *viz.*, 0.5926.

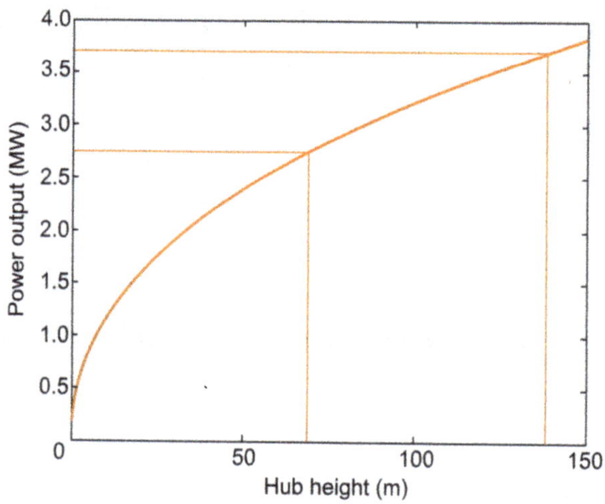

Fig. 5.12: Increase of power with hub height

The ideal power curve for a modern commercial wind turbine is that shown in Fig. 5.12. Below a threshold wind speed, the cut-in wind speed, the turbine is not operated because the losses in the drive-train and power converter are greater than the power extracted from the wind. Above the cut-in wind speed, the power generated is increased with wind speed until a second threshold, the rated wind speed, is reached. Power limiting is then undertaken to maintain a constant level of generation until a third threshold, the cut-out wind speed, is reached. Above the cut-out wind speed, the loads during operation would be unsustainable and the wind turbine is shut down. Between rated and cut-out wind speeds is the above-rated operating region. Very roughly, a wind turbine, well matched to site conditions, operates 25% of the time in below rated, 25% in the region of the rated and 25% in the above-rated wind speed conditions. It is non-operational, because the wind speed is either below cut-in or cut-out, for the other 25% of the time (Leithead 2007).

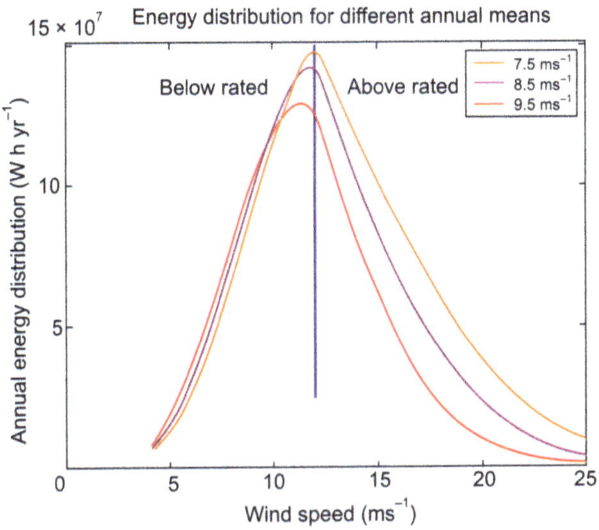

Fig. 5.13: Energy distribution with differing wind speeds
Energy distribution with differing wind speeds
(Courtesy: Leithead 2007)

5.3 Energy from Biomass

It is well known that organic matter can be used to provide heat, make fuels, chemicals and other products, and generate electricity. Wood, the largest source of bio energy, has been used to provide heat for thousands of years. But there are many other types of biomass—such as plants, residue from agriculture or forestry, and the organic component of municipal and industrial wastes—can now be used to produce fuels, chemicals and power (Fig. 5.14). In the future, biomass resources may be replenished through the cultivation of energy crops, such as fast-growing trees and grasses, called biomass feedstock. The degree of complexity and feasibility of biomass conversion technology depends on the nature of the feedstock from which we start (Congressional Record, 2005).

Fig. 5.14: Various forms of Energy yielding Biomass Various forms of Energy yielding Biomass (Composed and reconstructed from Google Image)

Unlike other renewable energy sources, biomass can also be converted directly into liquid fuels for transportation. The two most common biofuels are ethanol and biodiesel. Ethanol, an alcohol, is made by fermenting any biomass high in carbohydrates, like corn, through a process similar to brewing beer. It is generally used as a fuel additive to cut down a vehicle's carbon monoxide and other smog-causing emissions. Biodiesel, an ester, is made using vegetable oils, animal fats, algae, or even recycled cooking greases. It can be used as a diesel additive to reduce vehicle emissions or in its pure form to fuel a vehicle.

Fuel oil can be obtained by converting biomass chemically with the application of heat. Thereafter, the fuel oil can be burnt to generate electricity. Alternatively, steam can be produced for electricity production by directly burning biomass. The paper industry routinely feeds wood scrap into boilers to produce steam. Cold fire power plants also use biomass in high energy efficiency boilers as a supplementary energy source and in this way reduce emissions substantially.

Even gas can be produced from biomass for generating electricity. Gasification systems use high temperatures to convert biomass into a gas (a mixture of hydrogen, carbon monoxide, and methane). Certain microorganisms act on biomass like cow dung, wastewater from dairy and food processes, slaughterhouse waste, *etc*. and convert the biomass into biogas (containing methane and CO_2) and sludge through anaerobic digestion. The decay of biomass in landfills also produces methane that can be burned in a boiler to produce steam for electricity generation or for industrial processes.

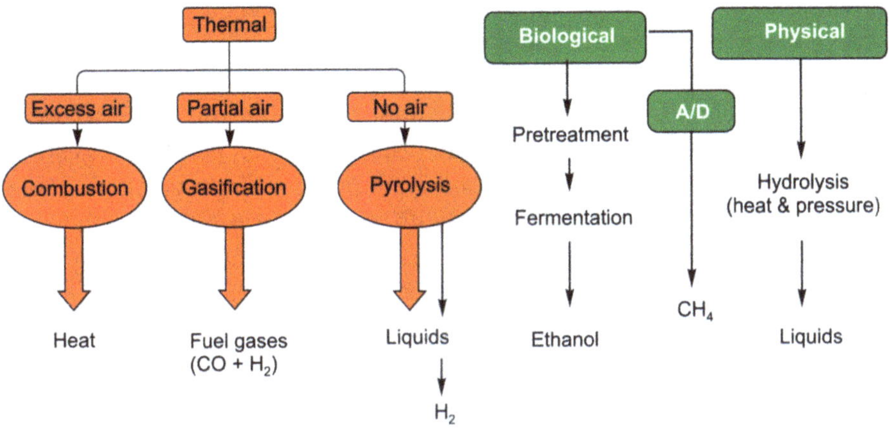

Fig. 5.15: Biomass energy pathways

Biomass gasification essentially breaks down wood and/or green waste into a synthetic gas composed mainly of hydrogen and carbon monoxide (Fig. 5.15). Controlling proportions of biomass, oxygen and steam within the gasifier yields various ways of gasification. Systems like fixed-bed, fluidized-bed and entrained-flow are also available for controlling the flow of biomass during gasification.

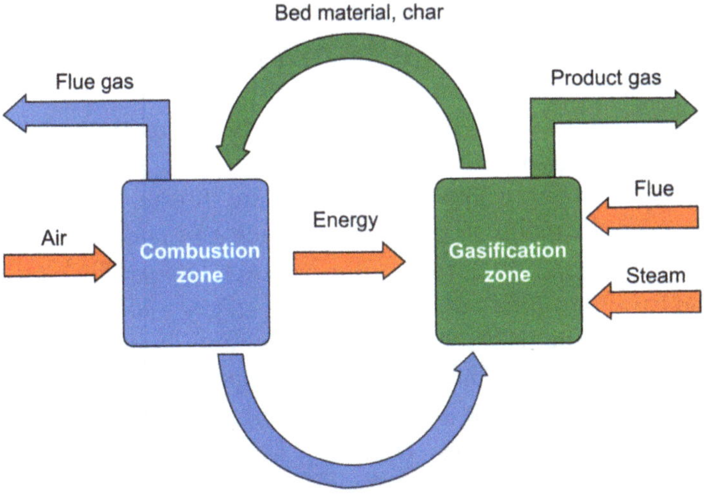

Fig. 5.16: Basic layout of the gasification process

5.3.1 Biogas Plant Designs

Anaerobic digestion is a process in which microorganisms break down biodegradable material in the absence of oxygen. The process is widely used to treat wastewater sludges and organic wastes because it provides volume and mass reduction of the input material. As part of an integrated waste management system, anaerobic digestion reduces the emission of landfill gas into the atmosphere. Anaerobic digestion is a renewable energy source because the process produces a methane and carbon dioxide rich biogas suitable for energy production helping replace fossil fuels. Also, the nutrient-rich solids left after digestion can be used as fertilizer.

The main stages of anaerobic digestion are:

- Hydrolysis
- Acidogenesis
- Acetogenesis
- Methanogenesis

Since biomass is made up of large organic molecules, bacteria in anaerobic mode must be able to first break these organic chains into smaller chains. The constituent parts and monomers, essentially sugars, are easily available to other bacteria. This process is called hydrolysis and constitutes the first step. Acetate and hydrogen produced in these first stages are used directly by the methanogens. The volatile fatty acids with a chain length exceeding those of the acetates are catabolised into compounds directly consumed by the methanogens.

During acidogenesis, there is a further breakdown of the remaining constituents by fermentative or acidogenic bacteria. In the process, Volatile Fatty Acids (VFAs), ammonia, carbon dioxide and hydrogen sulphide are the main byproducts. This third stage is technically termed as acidogenesis where simple molecules are further digested by the acetogens to yield acetic acid, carbon dioxide and hydrogen.

The fourth stage constitutes the process of methanogenesis–the utilization of the products of the preceding stages and their subsequent conversion into methane, carbon dioxide and water–the dominant components of biogas. The digestates comprise of mainly the non-digestible material that the microbes do not consume as well as dead bacterial remains. Methanogenesis is pH sensitive with the ideal range between 6.5 and 8.

A simplified generic chemical equation for the overall processes outlined above is as follows:

$$C_6H_{12}O_6 \rightarrow 3CO_2 + 3CH_4$$

The biodigester is a physical structure, commonly known as the biogas plant. Since various chemical and microbiological reactions take place in the biodigester, it is also known as bio-reactor or anaerobic reactor. The main function of this structure is to provide anaerobic condition within it. As a chamber, it should be air and water tight. It can be made of various construction materials and in different shape and size. Construction of this structure forms a major part of the investment cost. Some of the commonly used designs are discussed below.

Floating drum digester. Biogas technology in India started as early as 1937. Later in 1956, Jashu Bhai Patel developed the first gobar gas plant which is a floating drum plant (Fig. 5.16a). Subsequently, this design was approved by the Khadi and Village Industries Commission of India and soon became a benchmark design worldwide. The digester chamber has brick masonry in cement mortar with a steel drum placed atop the digester to collect the biogas. Subsequently, the floating drum plant type became less used with the introduction of the fixed dome model plant developed by the Chinese–this later design was easy to maintain and was more cost effective.

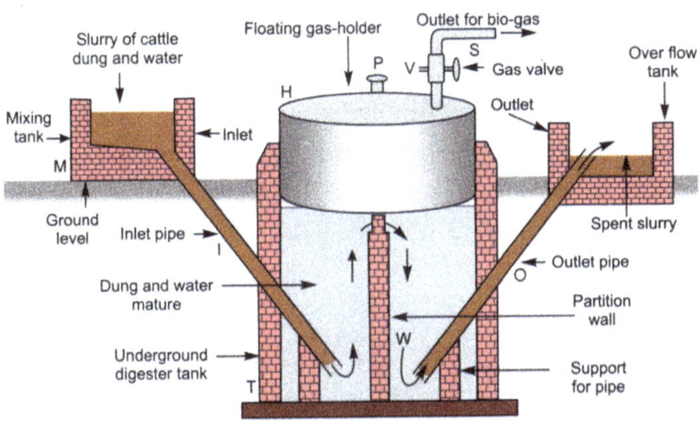

Fig. 5.16a: Definition sketch of a Floating Drum Biogas plant (Courtesy: http://www.nathmotors.com/)

Fixed dome digester. The fixed dome Chinese model was fabricated in China way back in 1936. It also comprises of an underground brick masonry fermentation chamber with a dome atop for gas storage–the fermentation chamber and the gas holder are essentially two components of the same unit (Fig. 5.17). By eliminating the costlier mild steel gas holder, the problem of corrosion is also ameliorated–the digester easily has a lifespan ranging from 20 to 50 years.

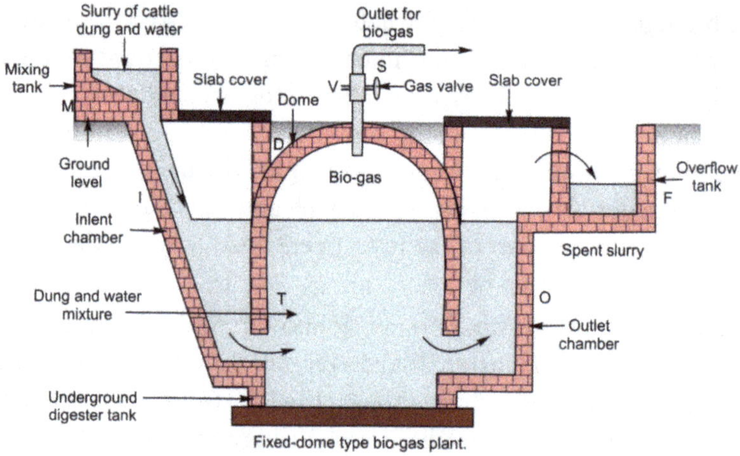

Fig. 5.17: A fixed dome digester (Courtesy: http://www.nathmotors.com/)

Deenbandhu model. In the famous Deenbandhu model developed in 1984, the investment cost of the preceding digesters was scaled down further. This was done under the aegis of Action for Food Production (AFPRO), New Delhi. The model was also cheaper than the popular people's model (the Janata model), essentially the first fixed dome digester plant based on the Chinese design. The Deenbandhu model was also fabricated entirely with brick masonry and has a spherical gas holder atop. The bottom, however, is concave (Fig. 5.18). This design is 45% cheaper as compared to the KVIC plant.

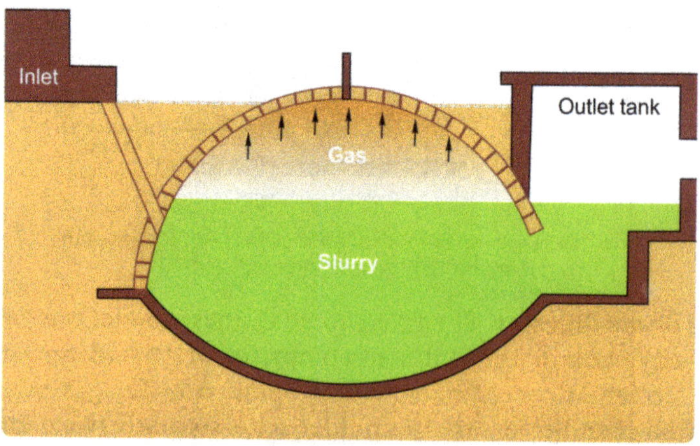

Fig. 5.18: Definition sketch of the common Indian Deenbandhu Biogas plant (Courtesy: http://grassrootsindia.com)

To summarize, the following considerations are of prime importance for the selection of a biogas plant.

1. Economic considerations: The plant should be cost effective in terms of cost per unit volume of the gas. Currently, with government subsidies it is expensive to an individual user but is modestly priced for a commune.
2. Design considerations: The design should be robust for operation and maintenance. This is especially important for developing countries with lower human development indices.
3. Local materials utilization: The use of locally sourced materials should be encouraged for all developing countries and particularly for those where transport infrastructure is not adequate.
4. Strength and durability: Strength and durability considerations should be factored into the calculation of the payback period. A biogas plant requires skilled labour which maybe scarce in the low HDI countries. Cheap, short-lived plants maybe initially cost effective but they are not amenable to reconstruction after its life cycle. Therefore, robust plants requiring a higher initial investment may prove to be cost worthy in the long run.
5. Suitability based on input material: Clearly the design should complement the type of bio-material input–when rice, straw and similar wastes are fed then a batch feeding design should be the preferred design rather than a semi-continuous feeding design.

In developing nations, people are looking to simple home based digestion systems affording cheap and low cost energy for some of the most basic needs–cooking and lighting homes. This need has been recognised by the United Nations Development Programme (UNDP) which promotes decentralized energy supply sources. Since 1975, both China and India have benefited from government supported schemes for the installation of biogas plants for the purposes of cooking and lighting. Such projects in the developing world can also be funded through the United Nations Clean Development Mechanism once the stakeholder has demonstrated that the process entails reduced carbon emissions.

5.4 Rainwater Harvesting in Monsoon Driven Countries

Rainwater harvesting entails first the collection and then the filtering and storing of water from roof tops. Rainwater is also harvested from paved and unpaved areas. And with a fool-proof filtration system, the water can be made potable after treatment. Any unused water can be used for raising the water tables through a process of recharging of ground water aquifers.

For example the overuse of fresh ground water has massively depleted the ground water reserves in and around Auroville near Pondicherry (Fig. 5.19). This has prompted Aurovillians to find means to replenish the ground water table and rainwater harvesting is the most obvious choice.

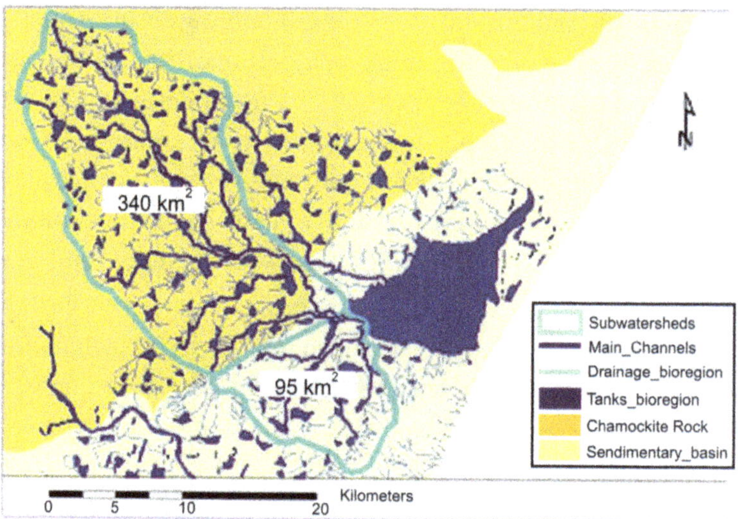

Fig. 5.19: Picture showing the Kaliveli sub-water-shed and other catchment areas which have been in use for above 1500 years[9]

5.4.1 Methods Adopted

The established methods include:

1. The direct method of in-site water harvesting with the use of terraces, underground tanks and dug up ponds.
2. The indirect method involving macro catchment harvesting with the use of recharge and dug wells, trenches and pits.

Fig. 5.20: A trench being dug in Auroville for rainwater harvesting[9]

Common adopted methods include afforestation of wasteland areas, building recharge structures, repair of the feeding channels to the tanks and awareness activities for farmers.

Fig. 5.21: Afforestation around an artificial basin created to collect rainwater[9]

In one of our previous chapters we have shown a pitched Keralan rooftop where rainwater is effectively harvested. This Indian practice is time tested wherein the building scale rainwater harvesting involves first collecting the water from building roofs and thence channeling it to a cistern which may well be free standing. Furthermore, each individual house therefore can have a self contained water supply system often with facilities embedded to filter and disinfect the water with traditional means. In better organized communes, clusters of buildings may well be connected to a larger storage area involving a rainwater collection network.

Fig. 5.22: Roof top rainwater harvesting in Tamil Nadu[9]

Rainwater harvesting techniques therefore are a panacea for rain starved districts of India leading to the sorely needed replenishment of the rapidly depleting ground water levels. There are of course other benefits:

1. Energy saving potential–a one meter rise in the water levels leads to a saving of 0.49 KWh of electricity.
2. Soil erosion and flood hazards are substantially reduced.

Fig. 5.23: Typical water harvesting Pond for Livestock[9]

5.5 Waste to Energy

Waste to Energy (WtE) is essentially waste treatment creating usable energy–electricity and heat from a source that would have been otherwise disposed off in a landfill. This energy recovery results in the production of synthetic fuels as well as methane, methanol and ethanol.

The key theme here is of course incineration and this maybe implemented without materials in energy recovery. However, this practice is not at all encouraged in the OECD countries where strict emission standards are enforced. Extant incinerators may well reduce the volume of the original waste up to 96% and this depends largely upon the composition and the recovery of metals and ash for recycling.

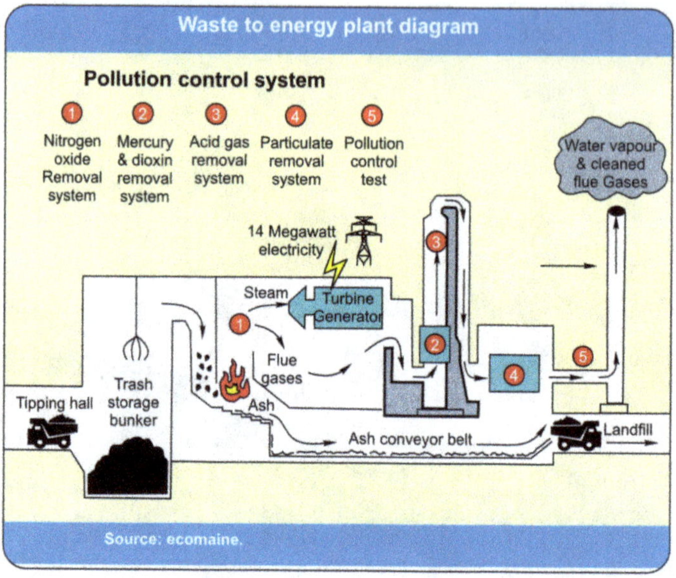

Fig. 5.24: Waste to energy pathways

Currently there are a number of new technologies yielding energy without direct combustion. Such processes separate the corrosive ash from the converted fuel. Here the higher combustion temperature in the boilers and IC engines or gas turbines efficiently converts the recovered energy into liquid and gaseous fuels.

Broadly classified, the thermal technologies include the processes of:

1. Gasification producing combustible gases, hydrogen and synthetic fuels.
2. Thermal depolarisation yielding synthetic crude oil–this can be further purified.
3. Pyrolysis yielding combustible chars, tar and bio-oils.
4. Plasma arc gasification or Plasma gasification process (PGP) yielding rich syngas with hydrogen and carbon monoxide used by fuel cells. They can also be used for generating electricity to drive plasma arches *etc*.

The non-thermal technologies include:

1. Anaerobic digestion with biogas rich in methane.
2. Fermentation production yielding ethanol, lactic acid and hydrogen.
3. A mechanical biological treatment which is followed by anaerobic digestion and finally through to a process entailing refuse derived fuel.

Finally we come to a discussion on quantifying the biomass fraction of waste fuels. Established methods include principles of radio carbon dating and the balance method. The latter employs existing data on the composition and operating conditions of the plant in question and thence calculates a probable result with a mathematical model. However, it must be borne in mind that even with a precise determination of the biomass fraction waste through carbon-14 dating, the calorific value still remains unquantified. The balance method therefore is the preferred method because it yields the ratio between biogenic and fossil fuel energy production along with the fossil mass and carbon fraction. To top it all, this latter method is easy to install and is cost effective.

References

1. http://www.daviddarling.info/encyclopedia/S/AE_sunspace.html
2. http://www.venhuizen-ww.com/docs/Rainwater%20catchment%20 proposal.pdf
3. http://www.erra.pk/Reports/Watsan/RWHConceptPaper.pdf
4. http://en.wikipedia.org/wiki/Waste-to-energy
5. http://www.envirothonpa.org/documents/16aBiomassBasics.pdf
6. http://www.ascentbioenergy.com/bio_gas_types.php
7. Nicholls, R., and Hall, K. (2006) The Green Building Bible. Green Building Press.
8. Lethead, W.E., (2007). Wind energy. Phil. Trans. Roy.Soc. A., 365, 957-970
9. Workshop on 'Essence of Energy Efficience-Auroville Case Study: Savitri Bhavan. Auroville 2009
10. Hanna, S. R., Briggs, G.A., Hosker, R.P. (1982). Handbook on Atmospheric Diffusion. Publisher: US Dept. of Energy
11. Congressional record (2005), V. 151 PT. 6. Edited by the Congress. Publisher: The Government of the United States of America.

6

Green Building Ratings

Chapter Overview

In order to provide guidance and assess the quality of sustainable buildings, different rating systems have been developed around the world. These systems help us compare one green building project with another, based on the numerous standard parameters that are identified by the system. Some of the popular rating systems are the LEED (U.S), BREEAM (U.K), GRIHA (India).

Before we begin to describe the various methods to rate green buildings, it is important that we understand that green themes are those concepts that have a low carbon footprint. This requires an understanding of Global Warming Potentials and Carbon Credits.

6.1 Global Warming Potentials, Carbon Credits and Clean Development Mechanism

Global warming potential (GWP) is a quantitative assessment of how much a given greenhouse gas contributes to global warming. The effect of GHGs is compared with that of carbondioxide which is the chosen standard. The GWP of carbondioxide is set to be 1 by convention. GWP depends on two important factors–the part of the spectrum that absorbs infrared radiation for a given species and its atmospheric lifetime. Greater the absorption and lifetime, greater is the GWP. The time interval associated with the measurement of GWP is crucial and is always expressed along with it.

Table 6.1: GWP Values of Different GHGs (Green House Gases) from the Intergovernmental Panel on Climate Change (IPCC 2007)

GWP values and life-times from IPCC (2007)	Lifetime (years)	GWP time horizon		
		20 years	100 years	500 years
Methane	12 (12)	72 (62)	25 (23)	7.6 (7)
Nitrous oxide	114 (114)	289 (275)	298 (296)	153 (156)
HFC-23 (hydro fluorocarbon)	270 (260)	12,000 (9400)	14,800 (12,000)	12,200 (10,000)
HFC-134a (hydrofluoro-carbon)	14 (13.8)	3,830 (3,300)	1,430 (1,300)	435 (400)
sulfur hexafluoride	3200 (3,200)	16,300 (15,100)	22,800 (22,200)	32,600 (32,400)

In order to compute the Global Warming Potential, the first step is to calculate the radiative forcing (RF) which is essentially the amount of energy per unit area over unit time absorbed by any greenhouse gas. Without this absorption, this green house gas would be lost to space. The Intergovernmental Panel on Climate Change (IPCC) lists values for the GWP. This is essentially the ratio of the time integrated RF from an instantaneous release of a kilogram of a trace substance relative to that of one kilogram of the reference gas which is carbon dioxide (CO_2):

$$GWP = \frac{\int_0^T A_x[g(t)]dt}{\int_0^T A_r[g(t)]dt}$$

Where T is the time horizon over which the calculation is considered; A_x is the radiative efficiency due to a unit increase in atmospheric abundance of the substance (Wm^{-2} kg^{-1}) and $[g(t)]$ is the time-dependent decay in abundance of the substance following an instantaneous release of it at time $t = 0$.

The denominator corresponds to the same quantity as in the numerator but instead with the reference gas-carbon dioxide. The radiative efficiencies A_x and A_r are no longer constant over time.

Carbon Credits are in essence emission permits created for a market to reduce green house emissions. A monetary value is assigned to the cost of polluting the air. One unit of carbon credit represents one tonne of carbon dioxide removed from the atmosphere or potentially saved from being emitted altogether. They are, in a sense, a method to certify countries promoting a reduction of GHG emissions. They are earned in the following manner:

1. Carbon sequestration through afforestation and plantation activities.
2. Implementing renewable energy technology to initiate carbon dioxide saving projects.

These credits are verifiable and are already proving to be a useful way of controlling emissions within free market economies. In this sense, carbon becomes a cost of business akin to other inputs such as labour and raw materials;

$$1 \text{ credit} = 1 \text{ tonne of } CO_2$$

6.1.1 CDM and other Mechanisms

The Kyoto Protocol provides a mechanism wherein developed countries with quantified emission limitations and stringent reduction commitments are allowed to acquire greenhouse gas reduction credits. These mechanisms are:

1. Under the Joint Implementation (JI), a developed country with high costs of domestic GHG reduction would have to set up projects in other developed countries with a relatively lower cost.

2. Under the Clean Development Mechanism (CDM), a developed country can take up a GHG reduction project in a developing country. Here, the project activities come with a much lower cost. The developed country is given credits for meeting its stipulated emission reduction targets, whilst the developing country receives the capital and the clean technology.
3. Under the International Emission Trading (IET), countries can trade freely within the international carbon credit market–those with surplus credits can sell them to countries with quantified emission limitation and reduction commitments under the Kyoto Protocol.

Although the Clean Development Mechanism is flexible, it is not free of criticisms. Establishing that CO_2 equivalence with a GHG is a complicated process. The first step therefore, is to determine whether a project has legitimately reduced emissions through the CDM mechanism. In this latter process participants submit a Designated Operational Entity (DOE). The Clean Development Mechanism Executive Board (CDM EB) has a panel in place which makes this system foolproof.

6.2 Indices for Labelling of Building Energy

Every building that is to be sold, must have a valid energy label. An energy label is a document stating the condition of the house in an energy context, and how it can be improved, *e.g.* by replacing windows, loft insulation, or replacing an old heating or an air-conditioning system.

The energy labels are separated into at least four categories:

1. Appliance details: This must be categorised separately for each appliance, its model and its material makeup.
2. Energy class: This is essentially a colour code (from A to G) that grades the appliance's electrical consumption.
3. Consumption, Efficiency and Capacity: This pertains to information on the appliance type.
4. Noise: Appliance generated noise in decibels.

In the example shown in Fig 6.1, houses are energy labeled on a scale from A-G in a way similar to that of electrical appliances. 'A' means that the building uses little energy, and 'G' means that the building consumes a lot of energy. Energy labels are generally prepared by specially trained energy consultants who have a thorough knowledge and understanding of this rating procedure.

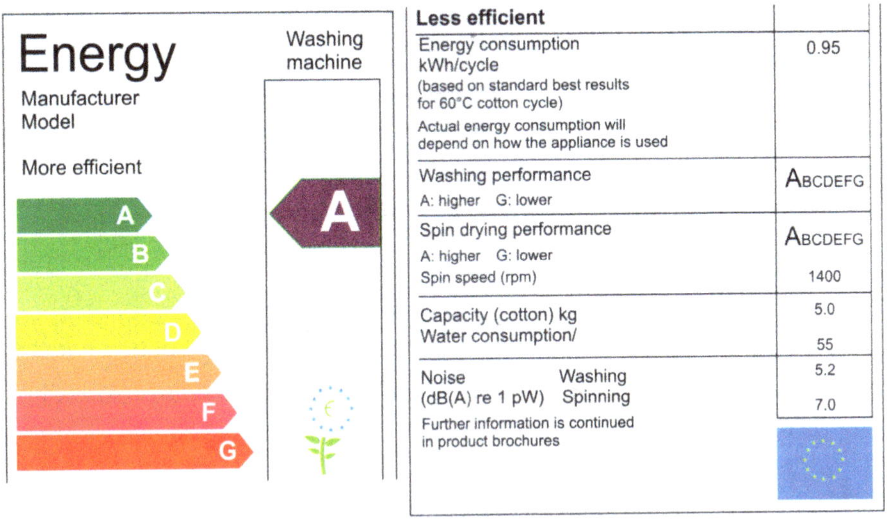

Fig 6.1: The european union energy label system for a washing machine (courtesy: http//www.swea.co.uk)

The first definitive step then is to identify the key parameters - compactness and insulation, the presence of thermal bridges, orientation and glazing quality, area, space heating efficiency, domestic hot water, lighting and ventilation and lastly, the use of renewable energies discussed earlier–solar domestic water heater, photovoltaic, *etc*.

6.3 Bureau of Energy Efficiency

Under the Energy Conservation Act (2001), Government of India established the Bureau of Energy Efficiency (BEE) on March 1, 2002. The Bureau develops policies and strategies with an overarching emphasis on self regulation and market principles. Within the remit of the Energy Conservation Act, the primary objective was reducing the energy intensity of the Indian economy. The Bureau claims that this is achievable through the active participation of all the stake holders. The end result would be an accelerated, yet sustained adoption of energy efficiency over all sectors.

The promotional activities, therefore, include dissemination of awareness, promoting conservation and organizing training of personnel in appropriate fields conversant with the efficient use of energy therein. Strengthening consultancy services and facilitating the implementation of pilot projects, the use of robust equipment devices and systems to conserve energy are also encouraged. The BEE also promotes innovative financing of projects to institutions, fostering energy efficiency and conservation and finally, international cooperation programs relating to such issues.

The BEE officially prioritizes the following objectives:

1. Provide policy and direction to all national energy conservation activities.
2. Coordinate policies and programs on the efficient use of energy.
3. Put procedures and systems in place to measure energy efficiency improvements.
4. Facilitate bi and multilateral leverage, including private sector support to implement the Energy Conservation Act (2001).
5. Robustly demonstrate energy efficiency delivery systems through public-private relationships.

6.4 Energy Auditing and Monitoring Software

Increasingly, Energy Management Software (EMS) are freely available (often real time through the web and with onsite display). Using them helps reducing costs and energy consumption in buildings and communities. These software essentially collect energy data for reporting, monitoring, and for meaningful engagement. Whilst Reporting includes a rigorous verification of energy data and sets reduction targets, Monitoring tracks consumption patterns to identify any cost saving opportunity. Lastly, engagement initiates a dialogue between managers and occupants, promoting conservation.

The main purpose of such software is to establish a link with the occupants' everyday choices for the home's energy consumption. If the consumption information are displayed on a real time basis, the dweller can immediately visualize the import of their action. Increasingly, such software offer advice on energy consumption initiatives and establishes a forum for like minded people to interact.

Letting occupants know their real-time consumption alone can be responsible for a 7% reduction in energy consumption [1]. Some of the popular commercial softwares are:

- Home Energy Analysis System–Marillack Pty Ltd
- Energy Lens–Energy Management Software–BizEE Software Ltd
- EHS Audit Management Software–2011 IHS, Inc.
- Recurve Inc.

EMS can also be an idiom for education–comprehensive dashboards can display energy information at homes through television. Alternatively, web browsers and clusters can be linked to a group to display this information. Some examples are given below:

1. **Energy Hub:** Google friendly spreadsheets and graphs of resources are offered by a high end energy dashboard known as the Energy Hub.

Fig. 6.2: Energy Hub device (Image courtesy–http://livemetering.blogspot.in/2009/04/10-monitoring-tools-bringing-smart.html)

2. **Tendril:** Tendril offers a combination of energy management services–wireless in-home energy display, cell phone applications, smart thermostats as well as an efficient web based energy portal. This allows users to identify and then regulate energy consumption.
3. **Agilewaves:** As the name connotes, this agile custom designed device can monitor gas, energy and water consumption real time using the World Wide Web. This needs an electrician for installation and is better suited to larger buildings. When used,

the device adjusts all smart appliances and can modulate light intensities in those and remotely modulate the central heating systems.

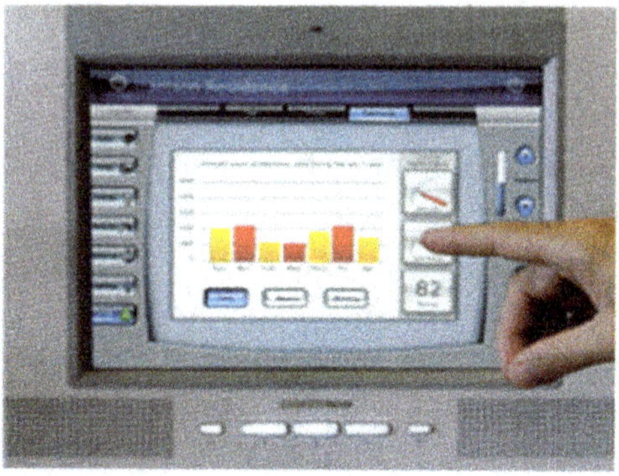

Fig. 6.3: A Screenshot of Agilewaves Software (Image courtesy–http://livemetering.blogspot.in/2009/04/10-monitoring-tools-bringing-smart.html)

Present day EMS are more sophisticated–they include various benchmarking tools tailor-made to inform energy consumption per unit area, weather normalization, advanced energy modelling algorithms that pick up any anomalous consumption patterns. It is clear that these advanced modules can greatly assist energy managers and auditors to identify saving opportunities.

The tedium of manual data entry is also overcome in this day and age with the use of computers–software are available for energy auditing efficiently and for monitoring power consumption in homes and offices. Increasingly smart meters yield breakup of energy analysis and even have the ability to control smart appliances.

Little drops add to an ocean–resource management using software yields enormous dividends. These tools help save time and cut down costs substantially making energy use management pleasurable.

6.5 Environmental Assessment Methods: BREEAM (Building Research Establishment Environmental Assessment Method)

BREEAM (BRE Environmental Assessment Method) is a widely used environmental assessment method for buildings delineating the best practice standards for sustainable design. In a sense, BREEAM measures the buildings environmental performance. It provides clients and property developers with appropriate market recognition for low and allowable environmental impact building's. It lends assurance that indeed the best practice has been incorporated into the building. It lends inspiration to seek solutions innovatively that minimize environmental impacts. BREEAM scoring system is tractable and transparent because it is supported by peer reviewed research. Without doubt, BREEAM has a benevolent influence on all aspects of building management.

BREEAM can be made country specific addressing many nuances of environmental issues with appropriate weightings and construction method details. It make references to local standards and building practised codes–indeed BREEAM UK, BREEAM Gulf and BREEAM Europe are already in place.

BREEAM awards credit in broad theme areas of Management, Health and Well-being, Energy, Transport, Water, Material and Waste among others. Weights are ascribed so that the credits can be added together to yield a resulting overall score. The BREEAM rated building is then graded on the scales testifying Pass, Good, Very Good, Excellent or Outstanding. A certificate is also awarded.

6.6 Energy Conservation Building Code

The Energy Conservation Building Code (ECBC) is part of the Energy Consumption Act of Government of India. The Bureau of Energy Efficiency placed under the Ministry of Power is aimed to stop wasteful use of energy and the energy consumption. Building codes were established for efficient energy management and conservation in the Indian built environment. In essence therefore, the ECBC benchmarks the minimum energy performance standards. The ECBC commercial buildings have proved to be cost effective with regulatory programs initiating energy savings.

ECBC compliance methods may involve the following methods:

1. Component–based: Requiring little energy expertise, it provides ample details on minimum performance requirements. On the flip side however, it is rather inflexible.
2. System–based: Here, some flexibility is afforded through a careful balance of mixing high efficiency components with low efficiency ones.
3. Whole building design analysis: This is flexible and sometimes exceeds energy efficiency requirements when benchmarked with a baseline building.

An example of a Building Envelope Design has been shown below.

Table 6.2: ECBC Building Envelope Design Strategy

ECBC Compliant Design Strategy for a Building			
Heat/Moisture Losses	Walls	Roof	Window
Minimize Conduction Losses	Use insulation with low U-value	Use insulation with low U-value	Use material with low U-factor
Minimize Convection Losses & Moisture Penetration	Reduce air leakage & use vapour barrier	Reduce air leakage & use vapour barrier	Use prefabricated windows and seal the joints between windows and walls.
Minimize Radiation Losses	Use light coloured coating with high reflectance	Use light coloured coating with high reflectance	Use glazing with low Solar Heat Gain Coefficient (SHGC)

(**Courtesy:** http//www.emt-india.net)

In summary, the ECBC is a boon to the individual and society at large. It affords tangible benefits to property owners and includes a reduction in energy costs without sacrificing comfort. It also entails reduced capital investments in energy supply systems and is characterized by an efficient management of resources.

6.7 Energy Rating of Appliances

From the preceding discussions, it was made clear that energy costs money, individually and communally. This calls for energy saving mechanisms with a concomitant reduction in energy bills and a much reduced reliance on conventional power stations for meeting our energy needs.

The Bureau of Energy Efficiency (BEE) introduced the 'National Energy Labeling Program'. This was mainly applicable for electrical appliances at home–the long shot of this labeling means that an individual can choose electrical appliances based on their performance.

The appliances are rated on a scale of 1 to 5–the most efficient product merits a 5 star rating.

Label Tubular Fluorescent Lamps

Star Rating	★	★★	★★★	★★★★	★★★★★
Lumens per watt at 0100 hrs. of use	< 61	< 61 & < 67	> = 67 & < 86	> = 86 & < 92	> = 92
Lumens per watt at 2000 hrs. of use	< 49	> = 52 & < 57	> = 52 & < 77	> = 77 & < 83	> = 83
Lumens per watt at 3500 hrs. of use	< 52	> = 49 & < 54	> = 54 & < 73	> = 73 & < 78	> = 78

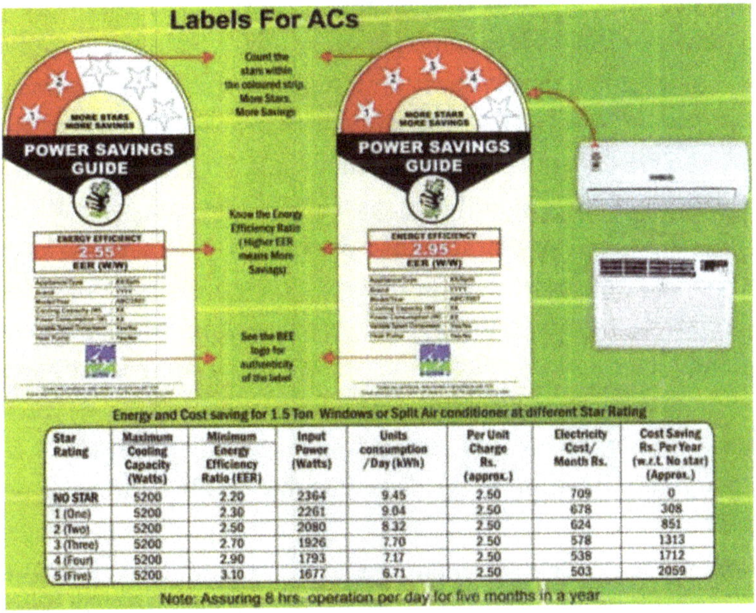

Fig 6.4: Energy Labeling System for TFLs and ACs by BEE
(Image Courtesy: http://www.beestarlabel.com/Content/Files/Schedule2_TFL.pdf and http://centralparkgurgaon.blogspot.in/2012/01/energy-saving-air-conditioning-tips.html)

Just as the national energy labeling program in India, the US Environmental Protection Agency developed the Energy Star program in 1982. Starred products agreed to money saving procedures amongst consumers and businesses whilst being kind on the environment. In addition, there have been other landmark advances in technology with the advent of LCD lighting and standby features in electrical products. A star rating entails strict energy efficiency guidelines.

Although India is definitely moving forward in energy conservation, the initiative will not be fully consummating until the minimum set standards are enforced. When unscrupulous manufacturers use it to their advantage, then the very purpose is lost.

6.8 Building Rating Systems

6.8.1 Leader in Energy and Environmental Design (LEED)

LEED is a well recognized green building certification–it authenticates a third party that a building was built aiming sustainability on all counts–energy and water conservation, CO_2 emissions reduction, proper environmental air quality within the built environment and resource management. This was developed by the US Green Building Council (USGBC) and provides owners a robust framework for the identification of green building parameters–design, operations and construction as well as maintenance paradigms with their solutions. It is flexible and transportable to all types of buildings–residential and commercial. It ramifies throughout the building life cycle–through design and construction stages to operation and maintenance, tenant fit out and increasing even retrofits. Need for neighbourhood development serves a local neighbourhood, extending its benefits laterally. LEED is a voluntary certification program extendable to all building types and life cycle phases.

Here, the points are awarded on a hundred point scale and weighted credits reflect their respective environmental impacts. Regional specific environmental issues are attributed through a 10 bonus credit system. A participating project must satisfy all essential prerequisites and a minimum number of points are required for certification by the Green Building Certification Institute (GBCI) for all registered projects.

Clearly, the LEED certification program not only encourages, but also actually accelerates the adoption of sustainable development practices globally. This is achieved through its fool-proof rating system described earlier.

Fig. 6.5: A sample of a LEED point system
(Image Courtesy: http://www.ecoarenas.com/LEED.html)

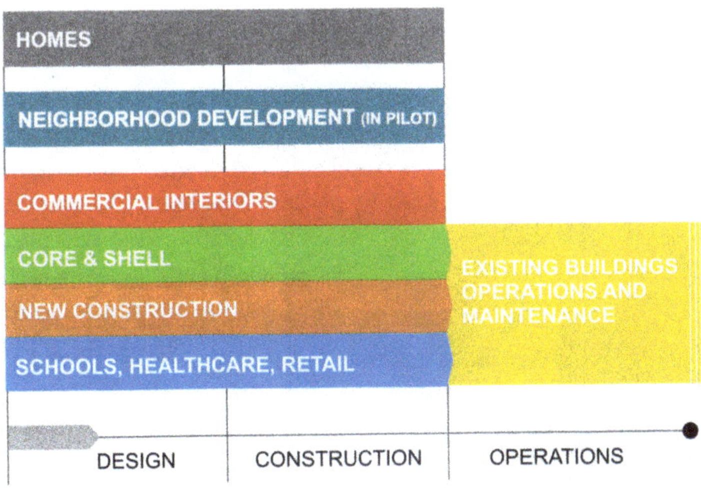

Fig. 6.6: Construction types covered under LEED
(Image Courtesy: http://www.ecoarenas.com/LEED.html)

Now we showcase some of the LEED rated buildings. But prior to that, we spell out the LEED certification levels:

1. Certified–40-49 Points
2. Silver–50-59 Points
3. Gold–60-79 Points
4. Platinum–80-110 Points

6.8.2 Green Rating for Integrated Habitat Assessment (GRIHA)

Now, we come to the Indian building sector and the associated ratings. The Indian building sector functioning within a thriving multicultural society is bound to be highly de-centralized–it is largely that. There are groups engaged in construction and equipment procedures as well as design and renovation–all may or may not work together. And this partly mellows down the integrated green design application process. This being the raison d'être, it is important to separate and demarcate the role of the different participants who ensure that the building in question is sustainable over its entire life cycle, leaving behind the minimal environmental footprint.

Fig. 6.7: Confederation of Indian Industry's Sohrabji Godrej Green Business Centre — Hyderabad, India (Image Courtesy: http://www.architectureweek.com/2004/1027/environment_2-1.html)

Fig. 6.8: Clinton Presidential Library — Little Rock, Arkansas (Platinum Rating) (Image Courtesy:http://www.easywaystogogreen.com/green-architecture/10-greenest-buildings-in-world/)

GRIHA is Sanskrit for 'abode'. The ancients in India had an intuitive idea about how buildings interacted with the environment. Even today, constructions should take into account building life cycles, resource consumption in the form of energy, water, materials and also factor in the ways in which the wastes are generated. GRIHA envisions a minimizing of the buildings' resource consumption, waste generation and ensures that its impact is within nationally acceptable standards.

The GRIHA rating system strikes a balance between well established practices both nationally and internationally with the guidelines updated every three years to factor in the latest innovations during this period.

The life cycle stages in GRIHA are:

1. Pre-construction stage: This includes the proximity to public transport, land and soil type, the property's location and the abundance of flora and fauna on the prescribed land prior to constructional activity, topography and land features and all aspects of landscaping.
2. Building planning and construction stages: This includes resource conservation and demand, resource utilization efficiencies, recovery and reuse. This also makes provisions for the occupant's overall health and well being. The resources factored in pertain to green cover, land, water, air and energy.

3. Building operation and maintenance stage: This includes issues related to operation and maintenance of buildings, the recording and monitoring of their energy consumption, occupant health and well being and finally any other issues effecting the local and global environment.

Although GRIHA has 34 criteria for their rating, only 4 are partly mandatory–all the rest are optional. Just like other rating systems, in GRIHA, each criterion has points associated with it. The levels of certifications range from 1 to 5 stars and are awarded on the basis of earned points–the minimum requirement for a certification is 50. A summary is sent out to members of the evaluation committee–well known experts involving the fields of landscape architecture, HVAC design, renewable sources of energy, day-lighting, water and waste management, *etc*. The rating is valid over a 5 year period from the date of commissioning of the building.

Within a 100 point system, the following marking system is adopted:

1. Certification–50 Points
2. 1 star–50-60 Points
3. 2 star–61-70 Points
4. 3 star–71-80 Points
5. 4 star–81-90 Points
6. 5 star–91-100 Points

Table 6.3: GRIHA Rating Scores (Courtesy: https//pcmcindia.gov.in/

Points scored	Rating
50–60	One star
61-70	Two star
71-80	Three star
81-90	Four star
91-100	Five star

Fig. 6.9: Centre for Environmental Sciences & Engineering Building at IIT Kanpur (Image Courtesy:http://epaper.timesofindia.com/Repository/ml.asp?Ref=VE9JTS8yMDA4LzEwLzEwI0FyMDA2MDE=&Mode=HTML&Locale=english-skin-custom)

Fig. 6.10: Suzlon One Earth Corporate Campus India–Pune (Courtesy: http//www. Grihaindia.org)

6.8.2.1 The Indian Green Building Council (IGBC)

The Indian Green Building Council (IGBC) was formed in 2001 as part of the Confederation of Indian Industry (CII). It developed the following rating systems (as of February 2010):

- LEED India for New Construction
- LEED India for Core and Shell
- IGBC Green Homes
- IGBC Green Factory Building

IGBC not only facilitates the green building movement, but also fosters the following services:

- Certification of Green Buildings in India
- IGBC Accredited Professional examination
- Green Building workshops & training programs
- Green Building missions
- Green Building Congress–India's flagship event on green buildings

References

1. http://en.wikipedia.org/wiki/Energy_management_software
2. http://gigaom.com/cleantech/10-energy-dashboards-for-your-home/
3. "The Prius Effect" by Elizabeth Dickinson
4. Comprehensive Dashboards to Enhance Employee Education and Awareness by Forward Energy Solutions, Inc.